Radio Astronomy Techniques.

By

R. N. BRACEWELL.

With 56 Figures.

ISBN 978-3-662-38653-8   ISBN 978-3-662-39512-7 (eBook)
DOI 10.1007/978-3-662-39512-7

I. Introduction.

1. Outline of the chapter. The first two parts of this chapter, concerned respectively with receivers and aerials, represent a selection from a vast field of radio technology. It is true that the material has been selected for its relevance to radio astronomy, but it remains a body of information on practice in radio technology which is current in connection with quite other fields than radio astronomy. It will be recognized as such by the expert in radio and will appear to him to be profoundly specialized in places and shallow to the point of omission in others. But parts II and III have not been written primarily for the expert in radio, they have been written for the physicist or astronomer, engaged in radio astronomy, who would benefit from a collected and orderly exposition of those parts of radio technique which are especially relevant to his work. Whilst the compilation consisted merely in selection, it is felt that the result will serve a valuable end, as the corpus from which the selection has been made is too extensive for one not specifically grounded in radio to use efficiently without guidance.

Very much of the material of prime importance for radio astronomy is available only in technical papers and some of the topics dealt with in parts II and III are not properly covered in the literature at all. The theory of the radiometer is one item which is still in an incomplete state and the effect of errors on the performance of aerials is another.

Part IV lays the groundwork for studying those parts of radio astronomy technique which have not been borrowed from existing practice but have been developed recently by radio scientists engaged in astronomical observation. The developments which have been called forth by the challenging new problems are impressive and are already being adopted in communications, navigation, radar and other established fields of application of radio science. Some refined concepts are essential to an understanding of the way in which an aerial explores a radiation field, concepts which would correctly be classified under classical optics. Indeed it would seem that recent advances in optics, especially in diffraction theory, have been taking place to a considerable degree at the hands of men interested in radio wavelengths. Part IV introduces the quantities *brightness*, *brightness temperature*, and *flux density* in terms of which observations are expressed, and sets up the *aerial smoothing equation*, an integral equation which relates them to *available power*, the quantity which is actually measurable with a receiver and aerial. From this equation a number of important theorems are deduced. The spectral sensitivity theorem relates the resolution of an aerial to its aperture distribution, in full detail; that is to say, not in terms of the aerial's ability to resolve a pair of point sources but in terms of the response of the aerial to spatial Fourier components of all spatial frequencies. The spectral sensitivity function is shown to be equal to the normalized complex autocorrelation function of the aperture distribution. The aerial cut-off theorem then shows that no aerial whose width is S wavelengths in a certain direction has any response to a sinusoidal brightness distribution with more than S cycles per radian in that direction. The discrete interval theorem, which follows as a consequence, sets a limit to the number of independent measurements which can be made in a finite area of sky, and shows that in conducting a survey it is sufficient to observe at discrete intervals of angle. All the results of part IV apply also to interferometers, which are defined as aerials having two or more well separated parts.

Part V describes the simple two-element interferometer, the radio analog of MICHELSON's stellar interferometer, and interprets its function as a determination of the complex visibility of a temporal interference fringe system. Under conditions of symmetry the complex visibility of the fringes observed with infinitesimal isotropic aerials is equal to the complex correlation between the field phasors at the positions of the aerials, and this in turn is equal to the normalized two-dimensional Fourier transform of the distribution of brightness temperature over the sky. This latter quantity, the brightness temperature spectrum, proves to be very suitable for describing interferometers and their behavior, and is to a good approximation measurable. Interferometers composed of extended aerials and sensitive over an extended frequency band are discussed in terms of the idealized isotropic monochromatic case, and are shown to measure smoothed values of the brightness temperature spectrum. Several extensions of the simple interferometer have been developed. Multiple element interferometers with sharp fringes permit quicker and simpler operation, phase-switching and lobe-sweeping introduce technical advantages, and asymmetrical arrangements have special applications. Closely related devices, which should not perhaps be considered interferometers, include phase-switched crossed linear arrays which attain greatly enhanced pencil-beam resolution in exchange for collecting area, and two-aerial systems not preserving phase which permit extreme separations of the aerials. An optical analog of this latter device, which introduces a radically new departure in optics, has been demonstrated by BROWN and TWISS in a measurement of the diameter of Sirius. It is probable that among the many striking advances in instruments for radio astronomy there are others which have significance for practical technique in other fields.

Part VI of this chapter goes into details of the measurement of radiation fields from celestial sources by means of both pencil beam aerials and interferometers. All the components of the often complex apparatus which is immersed in the field, the concepts necessary for considering its interaction with the field, and the procedures for absolute calibration, are touched on in the preceding parts, and clarify the observing procedures. Some very interesting considerations bearing on the analysis of the data close the chapter.

The fascinating subject of radio astronomy is in a vigorous stage of development. It has called forth striking advances in instruments for use at radio wavelengths and as its instrumental needs continue, continued rapid development may be expected to emerge from the profound joint resources of electromagnetic theory and established radio technology on which radio physicists draw. It is hoped that the present chapter on current techniques in radio astronomy is sufficiently fundamental and restrained in its inclusion of ephemeral matter to provide a good basis for following the expected developments of the next few years.

II. Receivers.

a) Principal receiver parameters.

2. Bandwith, gain, noise temperature, and time constant. A radio receiver is a non-linear device which converts high frequency electrical oscillations applied to its input terminals into an output electrical potential with certain convenient properties. The primary desideratum is sufficient power to drive some indicating instrument such as a recording ammeter, an oscillograph, a loudspeaker, a magnetic tape recorder, or typewriter. The output potential must also have a spectral distribution of power suitable to the needs of the indicating instrument. Furthermore one relies on the receiver to select the narrow band of high frequencies to which attention is to be directed. Among the necessary functions of a receiver, therefore, are (i) amplification, (ii) frequency conversion, (iii) tuning, (iv) selectivity.

The simplest receiver is one which puts out a steady potential depending on the

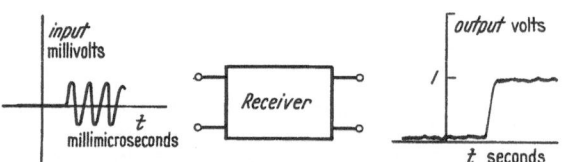

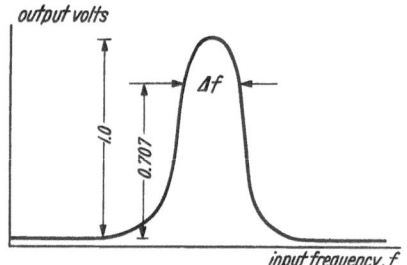

Fig. 1. Illustrating the amplifying and frequency converting properties of a receiver.

Fig. 2. Bandwidth of a receiver whose output voltage increases linearly with input voltage.

strength of the input oscillation. For example, the receiver in Fig. 1 puts out one volt with a monochromatic input signal of $10^{-3} \cos 10^{10}\,t$ volts.

In this case the voltage amplification is 10^3. The corresponding power ratio is referred to as the gain.

If input signals of other frequencies are considered the amplification becomes negligible outside a small range Δf centered on the nominal frequency f_0. Unless special effects, such as an unusually wide band, are wanted, the shape of the pass band will be approximately Gaussian since it is a product of the many similar characteristics associated with many successive stages of amplification.

The bandwidth Δf is generally defined as the difference between the frequencies at which the output voltage (or other output indication, such as meter deflection) falls to the value corresponding to a halved power input at the mid-frequency. Fig. 2 illustrates a case where output voltage increases linearly with input voltage. Bandwidths in practice lie mostly between 10^4 and 10^8 Hz.

As Figs. 1 and 2 show, the output of a receiver contains some random noise which is generated in the receiver itself, and except at the lower frequencies, around 20 MHz, it is desirable to keep the receiver noise as low as possible. The receiver noise is measured by the receiver noise temperature T_R, a quantity which would be zero for an ideal receiver. The ability of a receiver to detect weak signals depends, among other things, on receiver noise temperature, which is therefore one of the important parameters specifying a receiver.

It commonly happens that the non-linear stages of a receiver cannot be described by a simple mathematical law, and a consequence of this is that the gain depends on the input level. It is therefore necessary to calibrate receivers and in the process of doing so it is often possible to use a wideband noise power source and to include the aerial, feeder, and indicating device so that the monochromatic gain of the receiver alone loses primary significance. It is replaced by the gain calibration factor, measured in units of output indication per unit increment of input power. A typical value might be 1 millimeter per degree Kelvin, where the output is pen deflection on the chart of a recording meter, and the input power source is thermal or equivalent.

Further operations on the output signal of a receiver may be carried out by circuits which are best considered separately. For example the noise fluctuations shown in Fig. 1 may be reduced by passing the output signal through a low-pass filter which discriminates against the fluctuations. A direct consequence of this is to impede the rapidity of response to sudden changes. The smoothing circuit which does this is often referred to as an integrator. It is essentially a low-pass filter and in one simple form the response to a voltage impulse is an exponential decay with a time constant which may range from a fraction of a second to several seconds.

The integrating time τ, since it also affects the ability to detect weak signals, is an important basic parameter. The principal receiver parameters are thus (i) nominal frequency f_0 and its range, if adjustable, (ii) bandwidth Δf (and shape of band if significantly non-Gaussian), (iii) gain calibration factor, or detailed gain calibration if, as is common, the output indication is not a linear function of input power, (iv) noise temperature T_R, (v) integrating time τ (or impulse response of integrator). Precise meaning is given to Δf, τ and T_R in Sect. 8 and 10.

b) Monochromatic operation of a noise-free receiver.

In order to simplify the description of receiver operation it is convenient to ignore receiver noise. The block diagram of Fig. 3 is applicable to most receivers.

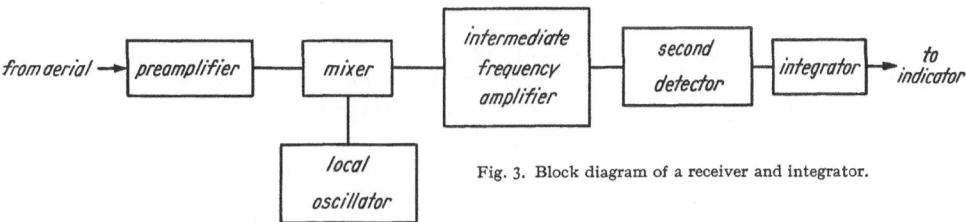

Fig. 3. Block diagram of a receiver and integrator.

It is also possible, by concentrating the amplification at the signal frequency, to omit the mixer and intermediate frequency amplifier, and it is advantageous to do this in the microwave range by cascading wide-band travelling wave tubes where large values of $\Delta f/f_0$, or special behavior of the receiver band pass characteristic, are required.

3. Preamplifier. A preamplifier means essentially an amplifier which amplifies at the original signal frequency, and may or may not be present in a receiver. If used, it may be situated away from the rest of the receiver. The main considerations in connection with a preamplifier are its effect on the noise temperature of the receiver, as discussed below, and the suppression of certain spurious responses. For frequencies greater than a slowly increasing limit, at present about 10000 MHz, it is an advantage to omit preamplification, since suitable amplifiers are only gradually being developed. There is a transition band of a few thousand MHz in which equally low noise temperatures are obtainable with or without pre-amplifiers (using triodes or travelling wave tubes), and at all lower frequencies triode or pentode preamplifiers are universal. If the attenuation between the aerial and the receiver is not negligible, preamplifiers must be reconsidered, and it has proved advantageous to use them at or near the focus of paraboloidal reflectors and at the extremities of extended interferometers (Sect. 60).

4. Mixer and local oscillator. The signal from the preamplifier and a strong signal from the local oscillator are combined in the mixer, a non-linear device such as a silicon crystal or suitably biased electron tube. Some energy now appears at the difference frequency, often 30 MHz, and is passed on to the intermediate frequency amplifier which accepts only a fixed band of frequencies centered on 30 MHz. Since the local oscillator frequency differs by 30 MHz from the nominal frequency f_0, any change in local oscillator frequency results in a change in f_0. This feature may be used for adjustment of f_0 but usually f_0 has to be kept fixed. The main requirement on a local oscillator is thus usually frequency stability. Other requirements are that it should give sufficient power at a steady level, be tunable as needed, and not absorb signal power. Provided the local oscillator power is strong, the envelope of the input signal is reproduced linearly.

5. Intermediate frequency amplifier. The intermediate frequency amplifier contributes all or most of the gain of the whole receiver and consequently must

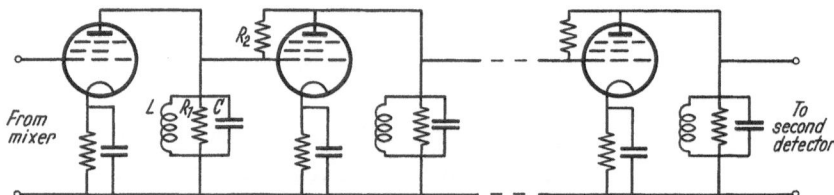

Fig. 4. An intermediate frequency amplifier.

meet the strictest demands on circuit design. Since the gain is typically one million, the minutest fraction of output power will cause instability if it leaks through to the input. For this reason intermediate frequency amplifiers have become highly developed along standardized lines. A great deal of skilled effort is needed to alter and readjust an existing design, hence the concentration on 30 MHz and other standard intermediate frequencies.

In Fig. 4 we see the basic pattern of an intermediate frequency amplifier[1]. A large number of amplifying stages are used, each with a selective tuned circuit LC. Since the effect of many tuned circuits in tandem is to produce bandwidths which are narrow, damping resistors R_1 may be provided. The negative feedback via R_2, if included, also tends to widen the band. The cathode resistor furnishes d.c. feedback which tends to stabilize the anode current and transconductance and should be chosen for the best time-stability. Inessential resistors, such as screen resistors, whose slow changes can influence the gain, should be omitted. Where the broadest bands are required, the frequencies to which the successive stages are tuned may be staggered.

Among the advantages of using a frequency change or superheterodyne receiver as described above are the ability to use a standard fixed-frequency intermediate frequency amplifier and the ability to adjust f_0 by a simple change in local oscillator frequency. Furthermore, the aerial structure is well adapted for picking up energy leaked from the receiver at the nominal frequency which makes a frequency change helpful if instability is to be avoided.

---

[1] G. E. VALLEY and H. WALLMAN: Vacuum tube amplifiers. New York-Toronto-London: McGraw-Hill 1948.

As a consequence of frequency changing the receiver is sensitive to two nominal frequencies each separated from the local oscillator frequency by 30 MHz. At the highest nominal frequencies there is little difference between what is received on these bands. However, the aerial pattern may be affected and for many purposes it is necessary to suppress one of the two. The one to be suppressed is called the image frequency. A selective preamplifier will usually suffice for image rejection. Otherwise an image rejection filter may be used. A superheterodyne receiver may also be sensitive to external signals at the intermediate frequency, and here again a preamplifier is effective.

6. Second detector. The second detector, so called because the mixer is regarded as the first detector, is a further non-linear frequency-changing device which performs the final step of converting the amplified signal to a steady output potential. It may consist of a diode rectifier (see Fig. 6) with suitable filtering to remove any traces of the intermediate frequency or its harmonics, or it may be some other standard detector utilizing a cathode follower or pentode. A linear detector is one whose output voltage or current is proportional to the amplitude of the voltage applied to it. The output of many detectors, however, is proportional to the square of the applied voltage amplitude, indeed at the lowest power levels virtually all detectors are square-law.

Before proceeding to the integrator which follows the second detector it is necessary to consider signals which are not monochromatic but are spread continuously over the spectrum.

c) Reception of a continuous spectrum.

7. Effect of non-zero bandwidth. Throughout the foregoing description of the parts of a receiver we have had in mind a monochromatic input signal and we have shown how it is converted to a steady output potential which measures its strength. However, the signals received from extra-terrestrial sources are not monochromatic; their energy is spread continuously over the spectrum. Let us consider such a signal as it passes through a receiver, and let us suppose that its spectral energy density is independent of frequency over the band of frequencies accepted by the receiver.

Fig. 5 (a) illustrates the time variation of electric field due to extra-terrestrial signals covering a wide spectrum as in Fig. 5 (b). The part of this field to which the receiver is sensitive is approximately monochromatic with frequency f_0 [Fig. 5 (c)] and has the spectral energy distribution shown in Fig. 5 (d). Because of beating between different components within the band, the waveform in (c) shows modulation at low frequencies which extend from zero to frequencies of the order of Δf. The important thing to notice is that the amplitude does not change much in times less than $(\Delta f)^{-1}$.

The output from the intermediate frequency amplifier, shown in (e), has the same envelope[1], but the mid-frequency has been shifted to 30 MHz as shown in (f). The second detector now suppresses or rectifies the negative half cycles and filters

---

[1] The probability of finding the envelope amplitude R between R and $R + dR$ is $p(R)\, dR$, where

$$p(R) = \frac{2R}{\langle R^2 \rangle}\, e^{-\frac{R^2}{\langle R^2 \rangle}}.$$

This is the Rayleigh distribution, and has the following properties: mean $= \frac{1}{2}\sqrt{\pi \langle R^2 \rangle}$, r.m.s. departure from mean $= \sqrt{(1 - \frac{1}{4}\pi)\langle R^2 \rangle} = 0.52 \times$ mean.

out higher frequency components to give an output (g) like the envelope of (e). This is the receiver output potential, which in the case of a monochromatic input signal was steady, but which now varies about a mean value. However the variations are no more rapid than the bandwidth Δf permits. For example,

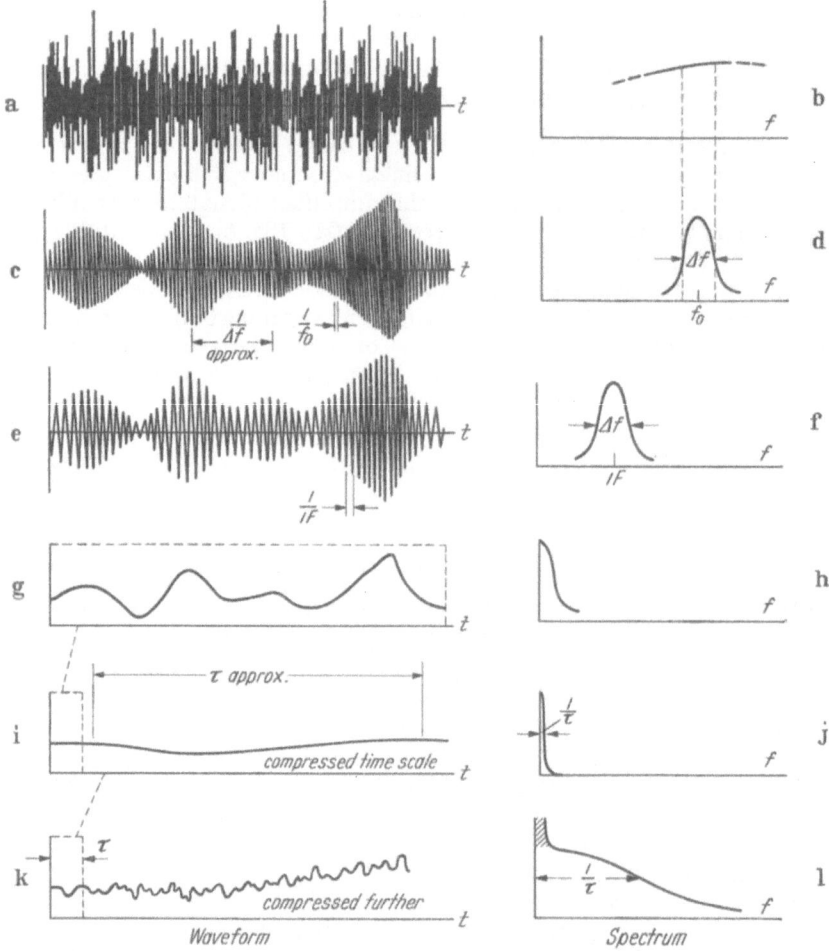

Fig. 5 a—l. Illustrating the changes in waveform and spectrum suffered by a signal with a continuous spectrum as it passes through a noise-free receiver.

if $\Delta f = 1$ MHz then the output cannot change much in times small compared with one microsecond.

The probability distribution of the receiver output voltage will depend on the detector law. If the detector is linear, the output voltage V will be the same as the envelope of (e) and thus have a Rayleigh distribution

$$p_1(V) = \frac{2V}{\langle V^2 \rangle}\, e^{-\frac{V^2}{\langle V^2 \rangle}};$$

more generally, the probability $p_2(V)\, dV$ of finding the output voltage between V and $V+dV$ is the same as the probability $p_1(R)$ of finding the input envelope between R and $R+dR$, i.e. $p_2(V)\, dV = p_1(R)\, dR$. For square-law detection

$V = R^2$ and $dV = 2R \, dR$; hence

$$p_2(V) = \frac{1}{\langle V \rangle} e^{-\frac{V}{\langle V \rangle}}.$$

This purely exponential one-parameter distribution has r.m.s. deviation from the mean equal to its mean value $\langle V \rangle$. Thus, as in the case of the linear detector, the r.m.s. deviation from the mean is proportional to the mean value.

To determine the input signal it is now only necessary to measure the mean value of the output.

d) Statistical limit to precision.

8. Standard deviation of a mean. We may assume the following theorem from statistical theory. If a large number N of independent variates are identically distributed (in almost any way) about a common mean M with r.m.s. deviation σ, then their mean is normally distributed about M with r.m.s. deviation $\sigma/N^{\frac{1}{2}}$. Regarding the receiver output as a succession of samples from a certain distribution we see that the uncertainty within which the mean can be determined depends on the length of time available for measurement and is proportional to the inverse square root of the number of effectively independent values assumed during the interval of measurement. Now independent values are separated by a time of the order of $(\Delta f)^{-1}$, so that about $\tau \, \Delta f$ independent values are assumed in an interval τ. Consequently, where the r.m.s. deviation of a single value from the mean is proportional to the mean, the uncertainty in the determination of the mean value is proportional to

$$\frac{\text{mean value}}{\sqrt{\tau \, \Delta f}},$$

which for $\tau = 1$ second and $\Delta f = 1$ MHz is equal to one thousandth of the mean value. This limit exists because of the random character of the signal to be measured, even in the absence of noise generated in the receiver.

One can assign precise definitions to Δf and τ to cover reception filters and smoothing filters with any frequency response. Let the power transfer characteristic of the reception filter be $R(f)$ and of the smoothing filter $S(f)$. The concept of "equivalent width" of a function proves to be basic, and is defined, for example for $S(f)$, by

$$W_S = \frac{\int\limits_{-\infty}^{\infty} S(f) \, df}{S(0)}.$$

We also require the following pentagram notation:

$$R \star R \equiv \int\limits_{-\infty}^{\infty} R(f') \, R(f' - f) \, df'.$$

We note that the output power spectrum from a square-law detector with input power spectrum $R(f)$ is

$$2R \star R + \left[\int\limits_{-\infty}^{\infty} R(f) \, df \right]^2 \delta(f),$$

where the first term enumerates the number of ways a difference frequency f can be found within the spectrum $R(f)$, and the second term is the rectified component. This spectrum, when further limited by the smoothing filter, becomes the power spectrum of the receiver output

$$S(f) \left\{ 2R \star R + \left[\int\limits_{-\infty}^{\infty} R(f) \, df \right]^2 \delta(f) \right\}.$$

The mean square fluctuation of the output is

$$2 \int_{-\infty}^{\infty} S(f)\,(R \star R)\,df = 2R \star R\,|_0 \int_{-\infty}^{\infty} S(f)\,df,$$

the mean value is

$$\sqrt{S(0)} \int_{-\infty}^{\infty} R(f)\,df,$$

and hence

$$\frac{\text{root mean square fluctuation}}{\text{mean value}} = \sqrt{\frac{2R \star R\,|_0 \int_{-\infty}^{\infty} S(f)\,df}{\int_{-\infty}^{\infty} R \star R\,df\,S(0)}}$$

$$= \sqrt{\frac{W_S}{\frac{1}{2} W_{R \star R}}}$$

$$= \frac{1}{\sqrt{\tau\,\Delta f}}.$$

The precise definitions will now agree with previous custom if we take

$$\tau = \frac{1}{W_S},$$

$$\Delta f = \tfrac{1}{2} W_{R \star R}.$$

The following tables, compiled with the assistance of Mr. R. Colvin, present a variety of cases for reference. For example, it is deducible from the tables that the r.m.s. fluctuation in the output of a receiver whose reception filter is Gaussian with half-power bandwidth B, and whose averaging is done entirely with a single RC circuit, is given by

$$\frac{\sqrt{\frac{\ln 2}{2\pi}}}{\sqrt{RCB}} \times \text{mean value} = \frac{0.56 \times \text{mean value}}{\sqrt{RCB}}.$$

Reception filter	$R(f)$	Δf
Rectangular pass band	$\begin{cases} 1, & f_0 - \tfrac{1}{2}\Delta < |f| < f_0 + \tfrac{1}{2}\Delta \\ 0, & \text{elsewhere} \end{cases}$	Δ
Two rectangular non-overlapping pass band	$\begin{cases} 1, & f_1 - \tfrac{1}{2}\Delta_1 < |f| < f_1 + \tfrac{1}{2}\Delta_1 \\ 1, & f_2 - \tfrac{1}{2}\Delta_2 < |f| < f_2 + \tfrac{1}{2}\Delta_2 \\ 0, & \text{elsewhere} \end{cases}$	$\Delta_1 + \Delta_2$
Triangular pass band	$\begin{cases} 1 - 2|f - f_0|/\Delta, & f_0 - \tfrac{1}{2}\Delta < |f| < f_0 + \tfrac{1}{2}\Delta \\ 0, & \text{elsewhere} \end{cases}$	$\tfrac{3}{4}\Delta$
Single tuned circuit	$[1 + (|f| - f_0)^2/\Delta^2]^{-1}$	$2\pi\Delta$
Two isolated tandem tuned circuits	$[1 + (|f| - f_0)^2/\Delta^2]^{-2}$	$\tfrac{4}{5}\pi\Delta$
Gaussian pass band[1]	$\exp[-(|f| - f_0)^2/2\Delta^2]$	$2\sqrt{\pi}\,\Delta$

Smoothing filter	$S(f)$	τ
Takes running means over time T	$(\pi T f)^{-2} \sin^2 \pi T f$	T
Single RC circuit	$[1 + (2\pi RC f)^2]^{-1}$	$2\,RC$
Two isolated tandem RC circuits	$[1 + (2\pi RC f^2)]^{-2}$	$4\,RC$
Critically damped RLC circuit	$[(R/2L)^2 + (2\pi f)^2]^{-2}$	$8\,L/R$
Rectangular pass band	$\begin{cases} 1, & |f| < f_0 \\ 0, & |f| > f_0 \end{cases}$	$1/2\,f_0$
Gaussian pass band	$\exp(-f^2/2f_0^2)$	$\sqrt{2\pi}\,f_0$

[1] The half-power bandwidth B is given by $B = \sqrt{8 \ln 2}\,\Delta$.

9. Integrator. A method of averaging the fluctuating output is to pass it through a low-pass filter, e.g. a simple combination of one resistor R and one capacitor C with a time constant RC. In this way the measurement of the $\tau \Delta f$ independent values is carried out automatically by the integrator, which gives out a potential approximately equal to the desired mean, but subject to small variations of order $(\tau \Delta f)^{-\frac{1}{2}} \times$ mean which assume independent values at intervals of order τ. Electronic counters also furnish a convenient method of averaging by performing and printing running sums over consecutive intervals.

Fig. 6 gives the basic circuit of a diode second detector and integrator. The diode capacitor takes up a positive charge such that the steady leakage through the shunt resistor is balanced by the replenishment during moments when the noise peaks of the input voltage are sufficiently positive to pass current through the diode. The time average of the diode capacitor voltage then appears across the integrating capacitor.

This final electrical signal is the one which drives the recorder and is shown in Fig. 5 (i) on a time scale which is compressed so as to exhibit just one cycle of fluctuation. In recording, the time scale would be even more compressed, as in Fig. 5 (k), which shows the actual record of the pen recorder indicating device.

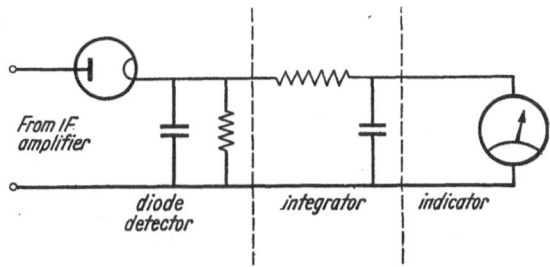

Fig. 6. Basic circuit of second detector and integrator in which $RC = \tau$

This device is often capable of following the most rapid variations passed by the integrator. Often, however, some further smoothing occurs in the recorder, which may for example be a moving coil galvanometer carrying a pen, and τ then assumes a larger effective value.

The constant of proportionality giving the amplitude of the final variations is calculable from the precise law of the second detector, the shape of the pass band of the intermediate frequency amplifier, and the shape of the impulse response of the integrator. Usually, however, the radio astronomer is unaware of the precise theoretical value but he does measure the strength of the variations and verifies that they are not unreasonably large. Many radio astronomical records reveal that the final minute recorded variations are non-Gaussian because of non-linear effects due to sticky ink, backlash, on-off servo-driven pens, etc. For these reasons, as well as the difficulty of determining the necessary parameters, the reasoning given above is most often resorted to for order-of-magnitude estimates.

10. Receiver noise. A noise-free receiver, such as has been assumed in the foregoing work can be approximated in practice by using sufficiently strong input signals. However, though the input power is reduced indefinitely, the output from the intermediate frequency amplifier will always give a positive indication due to noise generated internally in the receiver. Receiver noise is generated by statistical fluctuations in the rate of arrival of electrons at the anodes of the electron tubes induced by random times and velocities of departure of electrons from the cathode, thermal agitation in mixer crystals, and other causes. The first tube has by far the greatest effect since the noise contributions of later tubes are small compared with the already amplified signal.

4*

From a calibration such as that shown in Fig. 7 one can determine the gain g, the receiver noise output $g p_R$, and the amount of power p_R which, at the input, would account for the receiver noise output.

The quantity p_R is called the receiver noise power referred to the input terminals. It is a more suitable parameter than the noise output $g p_R$, which depends on gain, but is not free from dependence on Δf. To obtain a fundamental parameter specifying the noisiness of a receiver we refer p_R to the noise power available from a resistor at ambient temperature T_0 in the frequency band Δf, viz. $k T_0 \Delta f$, where k is BOLTZMANN's constant. The receiver noise temperature T_R is defined by

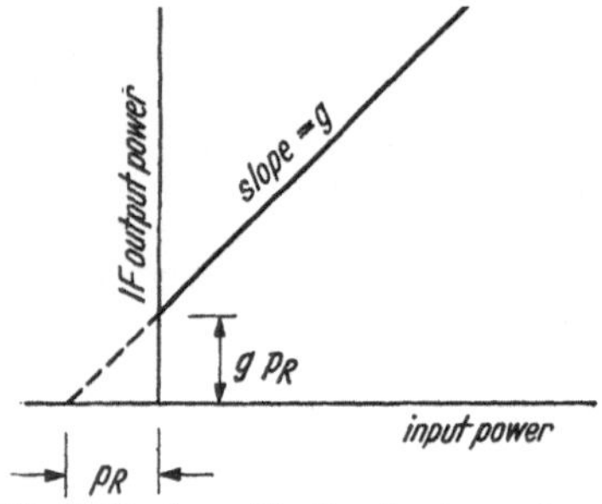

$$p_R = k T_R \Delta f$$

and the noise factor N is defined by

$$p_R = (N - 1) k T_0 \Delta f.$$

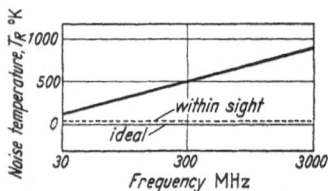

Fig. 7. Receiver calibration showing receiver noise.

The relation between the two parameters T_R and N is

$$T_R = (N - 1) T_0.$$

In an ideal receiver such that $p_R = 0$, T_R would be zero and N would be equal to unity irrespective of T_0. Confusion over the choice of T_0 makes T_R clearer than N when ambiguity must be avoided.

The expression $kT \Delta f$ for the power delivered by a resistor at temperature T into a matched circuit was established by NYQUIST[1] as follows. The thermal energy density per unit distance on a loss-free transmission line joining two resistors at temperature T is $2 k T \Delta f / c$; hence the power flow each way is $k T \Delta f$. To establish the energy density, suppose some energy is trapped by the sudden creation of two perfectly reflecting short circuits a distance l apart. The field must be expressible as a sum of natural modes at frequencies that are multiples of the fundamental $c/2l$. In any range Δf there are $2 l \Delta f / c$ such modes. Assigning energy kT to each mode in accordance with the principle of equipartition of energy, we find the stored energy $2 k T \Delta f \, l/c$.

At the highest frequencies (or at low temperatures), where kT exceeds the quantum energy hf, and all the modes therefore cannot be excited, allowance for the Boltzmann distribution of energy over the modes gives an average energy per mode of $hf/[\exp(hf/kT) - 1]$. The general expression for available power from a resistor is thus

$$\frac{h f \Delta f}{\exp(h f/kT) - 1},$$

Fig. 8. [Receiver noise temperatures currently attainable.

which is equal to $kT \Delta f$ in the Rayleigh-Jeans long-wavelength regime where $hf/kT \ll 1$, but is otherwise less. This expression essentially describes the Planckian black-body emission from a body whose radiation is confined to one polarization and constrained to flow in one dimension by a transmission line. The resistor is "black" if it is matched.

Fig. 8 shows noise temperatures that are readily available today, but values down to 20 or 30° K over the whole spectrum are in sight, and even lower values

[1] H. NYQUIST: Phys. Rev. 32, 110 (1928).

with special attention. Bloembergen's maser[1] demonstrated an amplifying principle capable of yielding very low noise by operation at liquid helium temperatures, and the closely related parametric amplifier, which can be operated at room temperature, is another promising device. There is reason to think[2] that a theoretical lower limit hf/k may exist. But this is extremely small and it is apparent that we are entering an era of essentially noise-free amplification in which the thresholds of sensitivity will no longer be set by limitations of receiver technique but by external phenomena. Cosmic noise at meter wavelengths and microwave thermal emission from atmospheric molecules are examples of such phenomena.

e) Calibration.

11. Thermal noise and shot noise. Because of the complexity of a receiver, overall calibration using input signals of known strength is a necessity and it is advantageous to make the calibration with noise signals.

A good deal of care has gone into this phase of radio astronomy technique since it is only by careful absolute measurements that results can be compared between different equipments and different frequencies. The possibility of using selected areas of sky as secondary standards has been much discussed but hitherto absolute calibration has not been avoidable. Aerial calibration also affects final results and will be discussed separately.

Noise of known strength can be had from resistors and from temperature-limited diodes for each of which reliable theory exists. Thus the noise power available in a band Δf from a resistor at temperature T is $kT\,\Delta f$, and the mean square noise current in a temperature limited diode in the frequency band Δf is $2eI\,\Delta f$, where I is the mean diode current and e is the charge of the electron. The first of these is deduced from thermodynamics (Sect. 10). The second is based on the assumption that the current I is composed of charge carriers of amount e which cross the interelectrode space with random times of arrival and departure. Each electron carries a current $e\delta(t-t_k)$ with a flat power spectral density e^2, or $2e^2$ if negative and positive frequencies are combined. The power spectra of the I/e electrons per second, arriving independently, are thus additive, with a resultant $2e^2(I/e) = 2eI$.

12. Resistive noise sources. Resistive noise sources have the advantage of simplicity and accuracy, when used near room temperature. The available power is minute, of the order of 10^{-15} watts, but suitable for many purposes. In some applications to very faint sources, resistors have been cooled by liquefied gases,

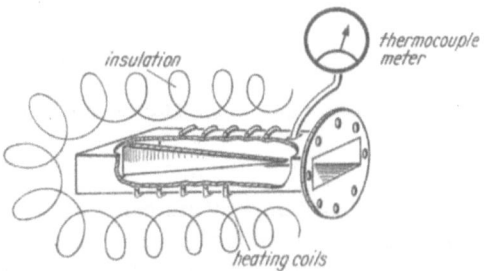

Fig. 9. A thermal noise source.

but more often hot resistors have been needed. The dependence of electrical resistance on temperature complicates the use of lumped resistors but it is convenient to use long lengths of slightly dissipative transmission lines. For example at wavelengths where waveguides are suitable a tapered length of resistive material is introduced into the guide (Fig. 9) in such a way as to absorb incident waves without appreciable reflection in the wavelength band of interest. The waveguide is then heated electrically, precautions being taken to ensure freedom

[1] N. Bloembergen: Phys. Rev. **104**, 324 (1956).
[2] K. Shimoda, H. Takahashi and C. H. Townes: J. Phys. Soc. Japan **12**, 686 (1957).

from temperature gradients, an exacting procedure if the highest precision is needed[1]. Finally the temperature is measured, usually with thermocouples.

By this means calibrating temperatures somewhat over 400° K are obtained. Higher temperatures have been attained by the use of a white-hot tungsten filament which is capable of rather higher but less accurately measurable temperatures.

13. Diode noise sources. Since calibration temperatures approaching one million degrees are wanted for some solar and galactic studies, something more intense than a thermal noise source is needed. For the longer wavelengths this is provided by the temperature limited diode.

By passing the mean square diode noise current $2eI\,\Delta f$ through a large resistor it is quite possible to produce noise powers approaching that available from a resistor at a million degrees. Fig. 10 illustrates a diode across which a potential difference high enough to draw the saturation current is maintained

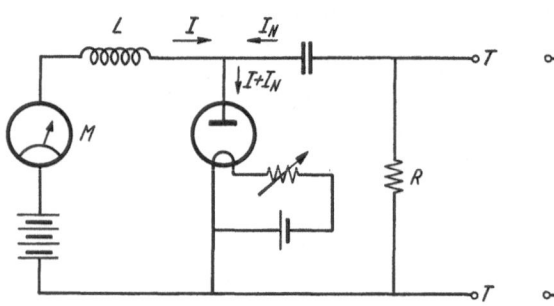

Fig. 10. A diode noise generator.

by a source shown as a battery. The diode current is $I+I_N$, I being the mean current and I_N the fluctuating part. A choke L prevents I_N from flowing in the battery circuit and a blocking condenser C ensures that the whole of I passes through the meter M. A resistor R defines the noise power which will be available from the terminals TT.

The noise power delivered to a load R_L is a maximum when the noise current i_N in the frequency band Δf is equally shared between R and R_L. Hence the available power is given by

$$\left(\tfrac{1}{2}i_N\right)^2 R = \frac{R\,e\,I\,\Delta f}{2}.$$

By writing

$$kT\,\Delta f = \frac{R\,e\,I\,\Delta f}{2}$$

we find for the equivalent source temperature T_s,

$$T_s = T_0 + \frac{e\,I\,R}{2k},$$

where the room-temperature contribution from the resistor R has been included.

In designing a diode noise source care is taken to make the blocking and choking elements adequate for the frequency band to be used, and adequate bypassing (not shown) for the meter is provided. A calculable correction may be allowed for. Too high a value of R will cause (i) leakage currents through conduction paths of unknown and perhaps variable resistance, (ii) fluctuations in potential comparable with the source potential, and (iii) problems of matching to loads. Stray capacity, including the interelectrode capacity, permits part of the diode current to bypass the resistor and therefore affects the accuracy of the calculated noise current. The source potential is not critical, but the current which heats the filament of the diode directly affects the noise output. It is necessary to stabilize this heating current but in addition it is usually made controllable as a simple method of adjusting the noise output.

---

[1] V. A. Hughes: Proc. Inst. Electr. Engrs. B **103**, 669 (1956).

The preceding considerations are all readily coped with up to frequencies of a few tens of MHz, beyond which the difficulties of measuring the resistance R, the current I, and above all eliminating stray reactance, cause the absolute accuracy to deteriorate. At higher frequencies, up to about 1000 MHz, diodes specially designed to fit without discontinuity into a coaxial line are available.

It is not necessary to include the resistance R but for practical reasons it is often desirable. For example if it is omitted small changes in R_L cause proportional changes in noise power whereas with R included, and R_L equal to R, poor adjustment of R_L has no effect to a first order, for $R = R_L$ maximizes the noise power delivered to R_L.

14. Gas discharge noise sources. In the microwave range gas discharge noise generators are available with noise temperatures above 10000° K. It appears that the velocity distribution of the free electrons in an ordinary mercury-filled fluorescent tube corresponds to a temperature of about 11000° K, and this temperature proves to be to a large extent independent of the discharge current and the gas pressure. To couple the thermal radiofrequency radiation resulting from collisions out of the discharge into a waveguide it is necessary to introduce the discharge tube into the waveguide in such a way that energy incident from a signal generator will be fully absorbed by the discharge. This presents no difficulty,

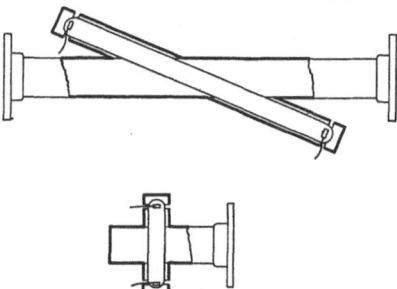

Fig. 11. Gas discharge noise generators in waveguide.

the arrangement introduced by MUMFORD[1] in his original work being to place the tube perpendicular to the axis of the guide. Matching arrangements are then necessary, but the scheme[2] illustrated in Fig. 11 (a) gives good matching over a wide frequency range. The perpendicular arrangement [Fig. 11 (b)] is suitable for precise measurement and also for fixed frequency operation, where it results in considerable space saving. At frequencies where waveguide is cumbersome, a suitable arrangement is to place the discharge tube on the axis of a special coaxial transmission line whose inner conductor is helical..

Because of their ready availability and range of sizes commercial mercury-vapour lamps with fluorescent coatings have been widely used. These tubes contain mostly argon, though nearly all the radiation emitted is from the small amount of mercury. Tubes are now

Gas	Noise temperature degrees K	Noise power db above $kT_0 \Delta f$
Mercury . .	11000	16
Argon . . .	11000	16
Neon . . .	18000	18

built especially for noise emission which contain pure argon and have clear glass tubes. Their noise temperature is almost the same but is more stable. Neon fillings are also obtainable with available noise 2 or 3 decibels higher than for argon. Careful work already done indicates that gas discharge tubes can perform as substandards to a precision of about one per cent. To achieve this accuracy in absolute level the very greatest care in calibration against a thermal source is at present required, but factory standardization is imminent. Meanwhile, the gas discharge source is very satisfactory to use, the absolute level being obtainable to about 10% from the accompanying table.

[1] W. W. MUMFORD: Bell. Syst. Techn. J. **28**, 608 (1949).
[2] H. JOHNSON and K. R. DEREMER: Proc. Inst. Radio Engrs. **39**, 908 (1951).

f) Sensitivity to weak signals.

15. The least detectable signal. It was shown earlier that

$$\Delta T \propto \frac{T}{\sqrt{\tau \Delta f}},$$

where ΔT is the uncertainty in the measurement of a temperature T, τ is the integrator time constant, and Δf is the overall receiver bandwidth, including aerial selectivity. If T is constant for the duration of many time constants, higher precision could have been achieved by the use of a longer time constant.

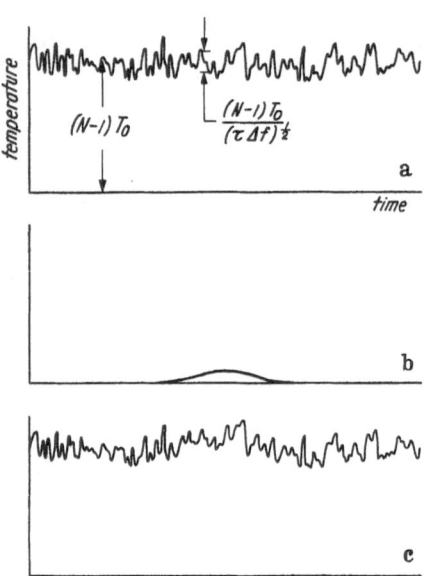

In practice one often gains this extra precision by scanning a pen record with the eye and smoothing it mentally, hence in discussing sensitivity to weak signals this aspect of τ must be borne in mind.

Suppose that the record shown in Fig. 12 (a) is obtained as an aerial sweeps over a part of the sky from which negligible radiation is received. Only receiver noise of amount T_R is recorded, and the fluctuations will be given by $T_R/\sqrt{\tau \Delta f}$. Now let the aerial sweep a part of the sky from which comes the faint signal shown in Fig. 12 (b); then the record obtained will be that shown in Fig. 12 (c). The question is how faint a signal will be detectable. In the figure the strength of the signal has been made roughly equal to the strength of the fluctuations and it will be seen that there is no doubt that this signal is detectable, for example by looking obliquely along the record.

Fig. 12 a—c. Showing how a signal (b) is readily detectable in equally strong noise (a) because of its duration.

Let the duration of the signal be $m\tau$. Then the value of T_R is determinate, over a length $m\tau$ of record, to $T_R/\sqrt{m\tau \Delta f}$, which is only a fraction of the strength of the recorded fluctuations. It is generally assumed that a signal of this strength is just detectable and hence that the sensitivity to weak signals is

$$\frac{T_R}{\sqrt{m\tau \Delta f}} = \frac{(N-1)\,T_0}{\sqrt{m\tau \Delta f}}.$$

If the efficiency of the aerial and feeder is η (Sect. 25), this expression becomes

$$\frac{\left(\dfrac{N}{\eta} - 1\right) T_0}{\sqrt{m\tau \Delta f}}.$$

The exact criterion of detectability would involve the profile of the signal to be detected, the percentage reliability of detection, the experience of the observer and his prior knowledge of the profile to be expected. Such a study would be important since some of the most interesting and controversial results of radio astronomy are drawn from faint signals on the limit of detectability, but it has not yet been undertaken. Instead, the effort so far has gone into building more and more sensitive equipment.

Where the utmost sensitivity is wanted superposition of small numbers of independent records may prove feasible as in the attempts to detect 21 cm radiation from the cluster of galaxies in Coma and thermal radiation from Mars.

If one considers using very large time constants of hours or days extreme requirements on stability of equipment are encountered. For example in Fig. 12 the recorded level was assumed constant on the average, but receiver output is proportional to receiver gain, a parameter which is notoriously prone to drift. In the example illustrated a signal amounting to about 15% of the receiver noise $(N-1)\,T_0$ was readily detectable, but we assumed that the receiver gain remained stable to much better than 15%. In the illustration the fluctuations were deliberately exaggerated and in practice one per cent of the mean noise would be more representative, implying gain stability to one part in 10^3 over times of the order of seconds. A faint signal of hours' duration, obtained say by tracking a planet, would require higher stability over a longer time. However, a stability of one part in 10^3 is itself no small achievement. Methods of alleviating the stability requirements will now be considered.

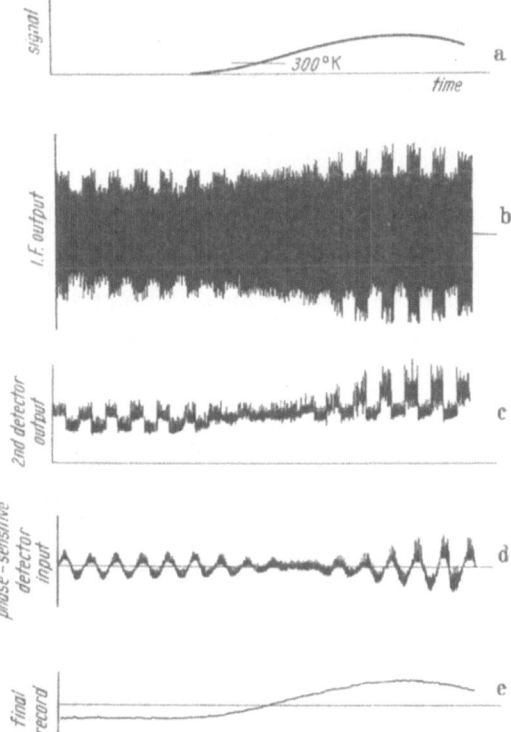

g) Comparsion systems.

16. Suppression of steady deflections. The intolerance of the direct system to changes in gain is essentially due to the largeness of the receiver noise relative to the signals to be detected. In the system due to DICKE[1], the steady compo-

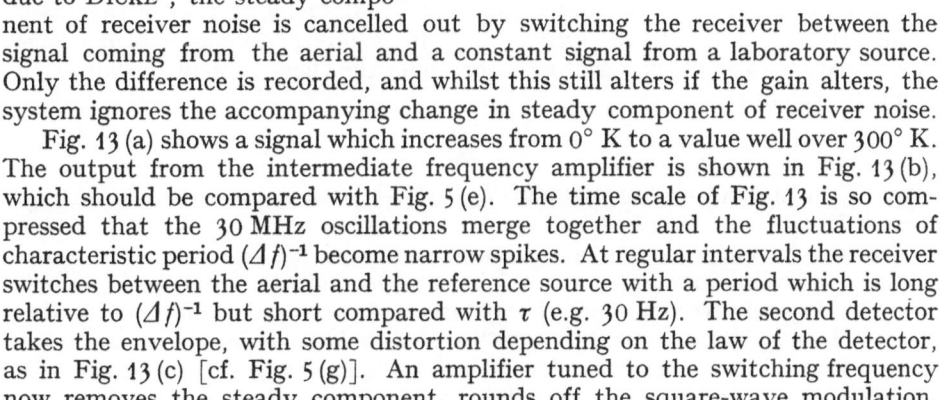

Fig. 13 a—e. Waveforms in a receiver incorporating the Dicke system of switching between the signal and a reference noise source.

nent of receiver noise is cancelled out by switching the receiver between the signal coming from the aerial and a constant signal from a laboratory source. Only the difference is recorded, and whilst this still alters if the gain alters, the system ignores the accompanying change in steady component of receiver noise.

Fig. 13 (a) shows a signal which increases from 0° K to a value well over 300° K. The output from the intermediate frequency amplifier is shown in Fig. 13 (b), which should be compared with Fig. 5 (e). The time scale of Fig. 13 is so compressed that the 30 MHz oscillations merge together and the fluctuations of characteristic period $(\Delta f)^{-1}$ become narrow spikes. At regular intervals the receiver switches between the aerial and the reference source with a period which is long relative to $(\Delta f)^{-1}$ but short compared with τ (e.g. 30 Hz). The second detector takes the envelope, with some distortion depending on the law of the detector, as in Fig. 13 (c) [cf. Fig. 5 (g)]. An amplifier tuned to the switching frequency now removes the steady component, rounds off the square-wave modulation,

[1] R. H. DICKE: Rev. Sci. Instrum. **17**, 268 (1946).

and reduces the amplitude and bandwidth of the noise [see Fig. 13 (d)]. The next step is to extract the amplitude of the switch-frequency oscillation, before finally smoothing.

Now it will have been noted that the phase of the oscillation in Figs. 13 (c) and (d) undergoes a reversal as the signal level passes through the reference

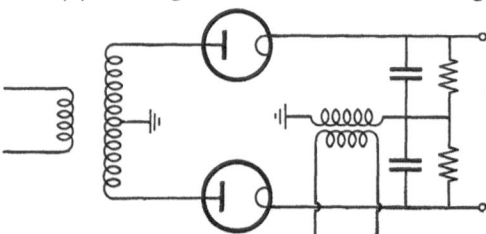

level, so that the final detector must be sensitive to phase. This is arranged by supplying the detector with a reference oscillation from the switch. A suitable circuit is shown in Fig. 14.

It will be perceived that the final deflection of the output indicator is a measure of the excess of the signal power over the thermal power

Fig. 14. A phase sensitive detector.

received from the reference source, i.e. it is proportional to $T - T_{\text{ref}}$; in the direct system the deflection was proportional to $T + T_R$, where T is the signal noise temperature, T_{ref} is the reference noise temperature, and T_R is the receiver noise temperature. Hence when T is approximately equal to T_{ref}, the sensitivity of

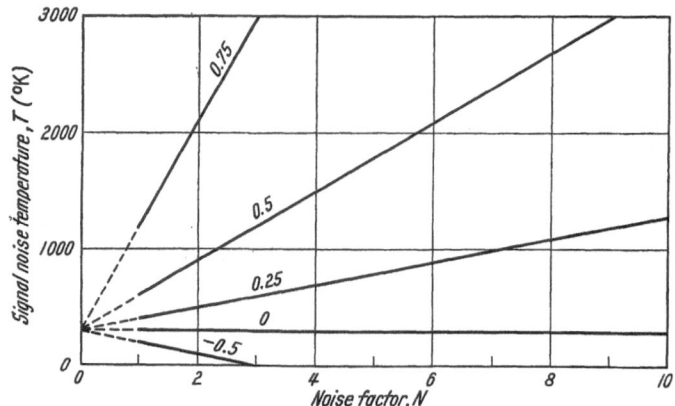

Fig. 15. The factors on the lines give the reduction in errors due to gain changes which results from adoption of the Dicke system.

the Dicke system to gain changes is negligible compared with that of the direct system. If a given fractional change in gain is interpreted as a signal level change then the ratio of the resulting error under the Dicke system to that using the direct system is

$$\frac{T - T_{\text{ref}}}{T + T_R}.$$

Taking T_{ref} equal to T_0 (290° K) and $T_R = (N-1) T_0$, we show this reduction factor in Fig. 15.

It is notable that when N is around 10 or more and the signal to be measured is in the range from room temperature right down to absolute zero, the advantages of the Dicke system are very great. The detection of microwave radiation from the. Sun and Moon by Dicke and Beringer[1] was carried out under these conditions

However, unless $T = T_{\text{ref}}$, the dependence on gain changes is serious enough to require great care in stabilizing the receiver. Typical measures which are

[1] R. H. Dicke and R. Beringer: Astrophys. Journ. 103, 375 (1946).

taken include temperature control of the room, stabilizing the high tension and filament supplies, and stabilizing the mean anode current of the electron tubes. A further scheme is to apply gated automatic gain control to the intermediate frequency amplifier in a way which tends to maintain the long-term average second detector output constant as sampled during the half cycles when the reference source is connected.

If T is expected to remain relatively steady, for example when a source is being tracked, there is an advantage in seeking a reference source of corresponding temperature, as in the examples below.

When T is high the advantage of the Dicke system falls off and in fact it has not been generally used in meter-wave studies of the intense radiation from the Sun and Galaxy.

17. Reference sources. DICKE used a rotating disc which mechanically inserted a dissipative wedge through a slot in waveguide at a frequency of 30 Hz. This method has since been widely used and has proved satisfactory. The motor which turns the disc also drives a generator which provides a phase reference signal for the phase sensitive detector. The reference temperature is room temperature, which of course may be subject to variation.

It is possible to move the absorption band of a piece of ferrite[1] placed in a waveguide onto and off the operating frequency f by varying a magnetic field maintained in the ferrite. This can be done at a frequency of hundreds of Hz, without sound or vibration. Again the reference temperature is room temperature, the temperature of the ferrite. In the interesting experiments with this technique reported by MAYER[2] the effect of aerial mismatch on accurate measurements was studied and an application was found for the ferrite isolator in ensuring tolerance to mismatch.

Gas discharges in coaxial line and in waveguide have also been tried as high temperature reference sources.

h) The Ryle and Vonberg system.

18. Towards independence from receiver gain change. The sensitivity to gain changes and the dependence on ambient temperature would be removed if the reference source were always at the temperature of the signal. This is feasible at wavelengths where noise diodes are practicable and was applied by RYLE and VONBERG[3,4] to the measurement of meter wavelength radiation from the Sun.

The temperature of the diode filament is varied so as to keep the noise power output equal to the signal. The receiver is called on merely to detect inequalities, and need have only a high gain, not necessarily a stable gain. There is thus a great advantage in principle over the Dicke system. The quantity measured is the diode current I, which directly indicates the signal power. Two characteristic features are (i) the thermal lag of the filament which sets a lower limit to τ, and (ii) the limitation to the range of noise power output available from the diode.

The Ryle-Vonberg system has worked satisfactorily in the meter-wavelength range, at frequencies around 18 MHz, and is being extended to microwavelengths. The physical embodiments of the principle differ in these three wavelength ranges, especially as regards the switch, the noise source, and the means of calibration,

[1] Proc. Inst. Radio Engrs. **44** (1956). This volume is devoted to the subject of ferrites.
[2] C. H. MAYER: J. Geophys. Res. **59**, 155 (1954).
[3] M. RYLE and D. D. VONBERG: Proc. Roy. Soc. Lond., Ser. A **193**, 98 (1948).
[4] K. E. MACHIN, M. RYLE and D. D. VONBERG: Proc. Inst. Electr. Engrs. III **99**, 127 (1952).

and are not developed to an equally high pitch of perfection throughout. However the basic advantages offered are great.

In the example of Fig. 16 the reference source is a gas discharge G in waveguide whose output is attenuated by a ferrite F subjected to a variable magnetic field. Reference noise levels from 18000° K down to 298° K are thus available. For an application to solar work, a more satisfactory range would be from 1800 to 30° K; hence one tenth of the power is coupled through a directional coupler D into a waveguide W which ideally would be terminated with an absorber near absolute zero. An accessible approximation is to terminate with an antenna A pointing to the celestial pole. A signal received in antenna B passes to the ferrite switch S, together with the power from the reference source. This switch, which is operated electrically several hundred times a second, alternately connects the signal and the reference to the receiver. Any inequality produces an output from the phase sensitive detector which alters the attenuation at F in such a way as to bring the reference level into equality with the signal. The current taken by the ferrite attenuator (indicated at M) is then recorded.

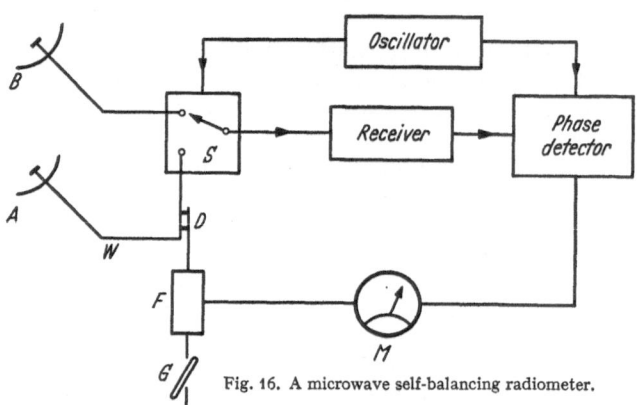

Fig. 16. A microwave self-balancing radiometer.

A modification for extending the range of operation below the minimum output of the noise generator was used by BROWN and HAZARD[1]. Locally generated noise was added to the signal noise and varied so as to maintain the sum constant.

III. Aerials.

Aerials are used in radio astronomy in much the same way as in other fields of radio, but with some differences. For example aerials may be used for transmitting or receiving but in radio astronomy, with the exception of astronomical radar, we are concerned only with receiving. For purposes of absolute measurement more attention must be paid to the determination of the effective collecting area of an aerial than is usually the case; polarization assumes a different importance; highly directional properties are called for; and methods of mounting and driving the aerial are quite distinctive. The various techniques practiced between 10^4 and 10^7 Hz do not concern us here.

The material which follows is a cross-section of aerial theory and practice selected for its appositeness to radio astronomy.

An aerial is a linear device which either (i) extracts power from a passing wave (often a plane wave) and delivers some of it to a pair of terminals or (ii) accepts power at its terminals and radiates some of it in all directions. The radiated field has a state of polarization which in general is elliptical and different in different directions; and a passing wave may always be split into two complementarily polarized components of which one is always fully rejected. The distribu-

[1] R. HANBURY BROWN and C. HAZARD: Monthly Notices Roy. Astronom. Soc. London **111**, 357 (1951).

tions of field and power flow around an aerial, and the currents flowing in it, are quite different in the receiving and transmitting cases. However, several reciprocal theorems closely bind the two uses, and we shall speak sometimes in terms of receiving, and sometimes in terms of transmitting, according to custom, even though the purpose is to receive.

a) Aerial parameters.

19. Radiation pattern. When power is fed into the terminals of an aerial currents are set up which produce magnetic and electric fields. Some of the energy of the fields flows into the conductors and surrounding ground and heats them, some ebbs and flows in the immediate vicinity of the aerial, and some is radiated away. Just as in the case of the diffraction fields produced by electric currents at optical frequencies, there is a division into Fresnel and Fraunhofer regions, only the latter being important in radio astronomy. In the Fraunhofer region, at great distances from the aerial, the flow of power is radially outwards and the electric and magnetic fields are transverse. The power per unit solid angle will however be a function of direction, $W(\alpha, \beta)$, where α is colatitude and β is longitude, referred to suitable axes in the aerial.

The function $W(\alpha, \beta)$, often pictured as a closed surface in space, or by means of representative cross-sections of this surface, is the radiation pattern. For a fuller description it would be necessary to add the state of polarization as a function of α and β and the departure of the isophase surface from spherical.

It is usual to normalize the radiation pattern in some way, especially by reducing the total radiated power or the maximum value of power per unit solid angle to unity.

20. Reception pattern. Consider a linearly polarized plane wave incident on an aerial from the direction (α, β), the plane of polarization making an angle $\psi(\alpha, \beta)$ with the meridian. Then the power delivered to a resistor connected across the aerial terminals will be a function of direction of arrival. This function of α and β is the reception pattern for the stated polarization and clearly other reception patterns exist for other conditions of polarization. For a full description one should include the phase of the current in the resistor under suitable assumptions of phase consistency of the plane waves from different directions.

21. Reciprocity theorems. The basic theorem stems back to work of HELMHOLTZ and RAYLEIGH and may be stated in terms of aerials and the intervening medium as follows. If a voltage generator G having zero internal impedance

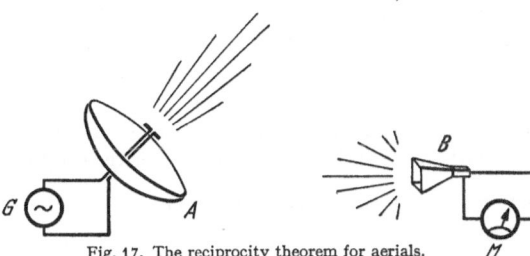

Fig. 17. The reciprocity theorem for aerials.

is applied to the terminals of an aerial A (Fig. 17) and produces a certain current in the zero-impedance meter M connected to the terminals of antenna B, then when the generator and meter are interchanged the current in the meter will be the same, under certain conditions. One of these conditions is linearity, both of the medium and of the elements of the antenna, another is freedom from the non-reciprocal effects associated with the presence of external magnetic fields, as in the ionosphere and in ferrites.

For theoretical purposes the reciprocity theorem is very important, as practical conditions for its applicability are readily realizable, though clearly also non-linearity and magnetic phenomena are commonly encountered in practice. Ground reflections, conductor loss, refraction, multiple paths, or transient conditions do not vitiate the reciprocity theorem.

A consequence of the basic theorem is that the radiation and reception patterns of an aerial are identical, provided the incident waves used for testing the reception pattern are suitably polarized. Another is that the impedance adjustment for maximum radiation is optimum for reception. These properties make it possible to test the reception of an aerial by using it to radiate, an arrangement which is often convenient.

In the usual pattern test arrangement the aerial under test and a second aerial are rotated relative to each other about an axis through the first. A generator is connected to one and the power received by the other is recorded. By the reciprocity theorem it will be immaterial which is used to receive, however the patterns so obtained refer not to the total power pattern $W(\alpha, \beta)$ but to a fraction of the power which is appropriately polarized. The remaining fraction, the cross-polarized component, is measured by rotating one aerial 90 degrees about the line joining the aerials and repeating.

22. Measurement of radiation patterns. The preceding discussion assumes a rigid aerial. When it is understood that the currents induced in the ground and objects surrounding an aerial are in effect part of the antenna, it will be seen that many rotatable and otherwise movable aerials are non-rigid. For this reason radiation patterns are usually virtually impossible to measure in their entirety.

When making measurements that are intended to apply to reception from celestial sources, it is necessary for the second aerial to be so far away that the path difference between any pair of rays joining the two aerials does not differ by more than a small fraction of a wavelength from the value it would have if the second aerial receded to infinity. Existing large aerials have already exceeded practical limits, but the partial check that is possible by flying a suitably oriented second aerial on a predetermined track at constant altitude through the distant field of the aerial under test has often been judged worthwhile. Direct observation of celestial sources such as the Sun and the strong discrete sources in Cassiopeia and Cygnus is helpful when the wavelength and beamwidth are such that only one source is important. While such observations may fall short of determining the directivity and details of the side radiation they may often have more pertinence than any other possible measurement.

23. Directivity. There is usually some direction which is of special interest in an aerial application such as a direction towards which most of the radiated power is concentrated. The directivity D of the aerial in this direction is defined by

$$D = \frac{\text{power radiated per unit solid angle in specified direction}}{\text{average power radiated per unit solid angle}}.$$

It is possible to calculate the directivity in terms of the radiation pattern. Thus if the specified direction is $\alpha = 0$, $\beta = 0$,

$$D = \frac{4\pi\, W(0, 0)}{\int\limits_0^\pi \int\limits_0^{2\pi} W(\alpha, \beta)\, \sin\alpha\, d\alpha\, d\beta}.$$

Since $W(\alpha, \beta)$ is difficult to measure, so likewise is D. However, in many cases it is possible to evaluate the integral to fair accuracy by combining one or two measured cross-sections of $W(\alpha, \beta)$, where it is large, with theoretical expectations.

24. Effective solid angle. The effective solid angle Ω is defined by

$$\Omega\, W(0,0) = \int\limits_{0}^{\pi} \int\limits_{0}^{2\pi} W(\alpha, \beta) \sin \alpha \, d\alpha \, d\beta,$$

whence

$$\Omega = \frac{4\pi}{D}.$$

For example if an aerial has a directivity of 1000 it has an effective solid angle of $4\pi/1000$ steradians or 41 square degrees.

25. Gain. If the power radiated by an aerial were radiated equally in all directions in an amount W per unit solid angle the total radiated power would be $4\pi W$. Now when power $4\pi W$ is supplied to the terminals of an actual aerial the power $W(0,0)$ radiated per unit solid angle in the direction $\alpha = 0$, $\beta = 0$ exceeds W by a factor g which is known as the gain. Thus

$$W(0,0) = g\, W.$$

Of the power $4\pi W$ being supplied to the terminals of an aerial let a fraction $\eta\, 4\pi W$ be radiated, the remainder being dissipated as heat in the conductors of the aerial and the surrounding ground. The quantity η is called the efficiency, and it relates the gain to the directivity; thus

$$g = \eta\, D.$$

The efficiency of an aerial is not easy to measure but it is usually high and so can be deduced from relatively rough estimates of the losses of power in the feeders and in the ground.

The gain defined here is sometimes referred to as gain relative to a loss-free isotropic source and so is not directly measurable. The measurable quantity is gain relative to some standard aerial, such as a half-wave dipole, whose directivity in equatorial directions is known from the theoretical radiation pattern to be 1.64. One takes the ratio of the input powers necessary to produce the same effect at a distance, and allows for the factor 1.64. It is still necessary to know the efficiency of the standard. The loss in the dipole itself is normally negligible, ground loss can be made negligible by raising the dipole sufficiently, and the feeder loss can be accurately measured. The most difficult matter is to ensure that no significant amount of power is carried away back along the feeder by unbalance currents. In measuring high gains it is customary to use a standard aerial of moderate gain which may itself previously have been calibrated against a dipole. Alternatively a horn of calculable directivity may be used (SCHELKU-NOFF[1]) or the two-aerial method of absolute gain measurement in terms of the inverse square law of spatial attenuation (SILVER[2]).

26. Effective area. When a receiving aerial is placed in a radiation field the power which it abstracts depends on the load impedance which is connected to its terminals. By adjustment of the impedance a maximum transfer of power to the load can be arranged, and this power is referred to as the available power of the antenna. The effective area A measures the ability of an aerial to abstract energy from the field of a traveling plane wave coming from a specified direction and to deliver it to the aerial terminals. Since only one polarization can be

[1] S.A. SCHELKUNOFF: Electromagnetic waves, Chap. 9. New York: Van Nostrand 1943.

[2] S. SILVER: Microwave antenna theory and design, Chap. 15. New York-Toronto-London: McGraw-Hill 1949.

accepted, the wave is taken to have that polarization. Then

$$A = \frac{\text{available power at aerial terminals}}{\text{power crossing unit area of wavefront}} \,.$$

It will be perceived from the reciprocity theorem that A is proportional to g, in fact

$$g = \frac{4\pi A}{\lambda^2} \,.$$

This relation is normally proved by calculating both g and A for some one loss-free aerial (e.g. see Schelkunoff and Friis[1]) but it is in a way unsatisfactory to have to appeal to a special case with its own inevitable approximations to establish a general result. The following explanation is taken from Pawsey and Bracewell[2]. An aerial has connected to its terminals a resistor at temperature T which has been adjusted for maximum power transfer from a passing wave to the resistor. The reciprocity theorem enables us to say that the adjustment is also correct for maximum power transfer from a voltage generator in series with the resistor to the aerial. A black body also at temperature T subtending a small solid angle Ω in a direction in which the gain is g intercepts power $g\,kT\,\Delta f\,\Omega/4\pi$. Using Planck's formula for black body radiation the power delivered to the resistor is $\frac{1}{2}(2kT\,\Delta f/\lambda^2)\,\Omega A$, the factor $\frac{1}{2}$ resulting from the reception of one component of polarization. Under conditions of thermal equilibrium the principle of detailed balancing requires these two powers to be equal. We then find that $g = 4\pi A/\lambda^2$ without appeal to details of any one aerial.

b) Radiation patterns.

27. Fourier transform relation. One way of relating the physical structure of an aerial to its radiation pattern is to integrate the remote (and duly retarded) effects of its constituent current elements and space charges. Another way is to begin from the fields impressed on an infinite plane in or near the aerial and deduce the distant fields. For the purposes of the present section the latter approach is advantageous.

We know that twin plane waves equally and oppositely inclined to the normal to a plane as in Fig. 18(a) give rise to a cosinusoidal standing wave pattern as indicated by the parallel nodal lines of the figure. Conversely, such a standing wave pattern impressed on the infinite plane must launch two such plane waves into the semi-infinite space on the side opposite the sources of the field. When the nodal lines are closer together the pair of wave-normals are spread more widely apart as in Fig. 18(b), the sine of the angle ϑ between the wave normal and the $\eta\zeta$-plane being equal to the number of cycles of the standing wave pattern per free-space wavelength. With $\vartheta = 90°$ the two waves interfere with like nodes separated by the free-space wavelength λ; as ϑ approaches zero the nodes separate indefinitely. When the impressed field has more than one cycle per free-space wavelength no traveling waves are launched, the field merely evanesces as indicated in Fig. 18(c). The value of considering these simple distributions is that they constitute Fourier components of more general field distributions.

When there is periodic variation in two directions as in Fig. 18(d) each ray is further split into two, each making an angle φ with the $\xi\zeta$-plane. Each two-

[1] S.A. Schelkunoff and H.T. Friis: Antennas theory and practice, Chap. 6. New York: Wiley 1952.

[2] J.L. Pawsey and R.N. Bracewell: Radio Astronomy, Chap. 2. Oxford 1954.

dimensional Fourier component

$$\cos \frac{2\pi\xi}{\lambda_1} \cos \frac{2\pi\eta}{\lambda_2} \cos \omega t$$

of a general impressed field distribution gives rise to four waves for which

$$\sin \vartheta = \pm \frac{\lambda}{\lambda_1},$$

$$\sin \varphi = \pm \frac{\lambda}{\lambda_2},$$

and the resulting disturbance is therefore of the form of the real part of

$$\exp\left[\frac{i\,2\pi}{\lambda}(l\xi + m\eta + n\zeta - ct)\right],$$

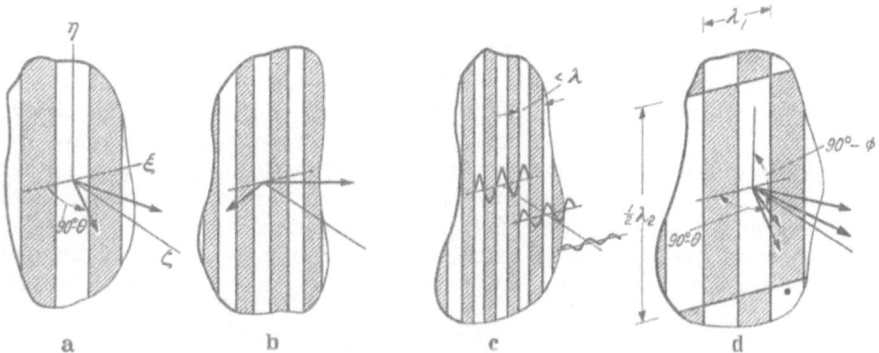

Fig. 18 a—d. Effects produced by a field distribution over an infinite plane aperture.

where l, m, n, the direction cosines of the wave normal, are given by

$$l = \pm \sin \vartheta,$$

$$m = \pm \sin \varphi$$

and

$$l^2 + m^2 + n^2 = 1.$$

The above wave function contains all the types of behavior illustrated in Fig. 18, including evanescent waves $(l^2 + m^2 > 1)$.

The basic point of this section is the Fourier transform relation between an aperture distribution of field and the radiation which it launches. Let an aperture distribution of finite dimensions be analyzed into all its doubly periodic components [see Fig. 18 (d)]; then the field at a point in the direction (ϑ, φ) at a great distance from the aperture is fixed by the strength of the Fourier component of period $\lambda \operatorname{cosec} \vartheta$ in the ξ direction and $\lambda \operatorname{cosec} \varphi$ in the η direction.

Let

$$P(l, m) \exp\left[\frac{i\,2\pi}{\lambda}(l\xi + m\eta + n\zeta - ct)\right] dl\,dm$$

be the complex expression for the field strength at (ξ, η, ζ) due to waves in the interval l to $l + dl$ and m to $m + dm$. Then the field $F(\xi/\lambda, \eta/\lambda)$ over the aperture plane $\zeta = 0$ is given by

$$F\left(\frac{\xi}{\lambda}, \frac{\eta}{\lambda}\right) = \int_{-\infty}^{\infty} \int_{-\infty}^{\infty} P(l, m) \exp\left[\frac{i\,2\pi}{\lambda}(l\xi + m\eta)\right] dl\,dm.$$

This is the standard form of the two-dimensional Fourier transform (Sned-don[1]); consequently the inverse relationship

$$P(l, m) = \int\limits_{-\infty}^{\infty} \int\limits_{-\infty}^{\infty} F\left(\frac{\xi}{\lambda}, \frac{\eta}{\lambda}\right) \exp\left[-\frac{i\,2\pi}{\lambda}\,(l\,\xi + m\,\eta)\right] d\left(\frac{\xi}{\lambda}\right) d\left(\frac{\eta}{\lambda}\right)$$

follows immediately.

From the function P, which will be referred to as the angular spectrum, following Booker and Clemmow[2], the radiation pattern can be determined. The power radiated per unit solid angle in the direction (l, m) will be proportional to

$$P\,P^* \sqrt{1 - l^2 - m^2},$$

since the element $dl\,dm$ subtends a solid angle

$$\frac{dl\,dm}{\sqrt{1 - l^2 - m^2}}.$$

When the aerial is highly directive, so that only values of l and m near zero are important, $P\,P^*$ represents the radiation pattern. Furthermore the aperture field distribution $F(\xi/\lambda, \eta/\lambda)$ then has negligible ζ-components, so that we arrive finally at the well-known statement that the field radiation pattern of an aerial is proportional to the two-dimensional Fourier transform of the transverse field distribution across its aperture. The preceding derivation will enable off-axis and polarization questions to be examined as necessary; meanwhile, taking the Fourier transform relation as a point of departure, one may immediately establish many aerial theorems.

The limits of integration $(-\infty$ to $+\infty)$ in all the double integrals which follow have been omitted, and we have taken $\sin\vartheta = \vartheta$ and $\sin\varphi = \varphi$.

28. Similarity theorem. By a change of variables,

$$P\left(\frac{\vartheta}{p}, \frac{\varphi}{q}\right) = p\,q \iint F\left(p\,\frac{\xi}{\lambda}, q\,\frac{\eta}{\lambda}\right) e^{-i\,2\pi\left(\vartheta\frac{\xi}{\lambda} + \varphi\frac{\eta}{\lambda}\right)} d\left(\frac{\xi}{\lambda}\right) d\left(\frac{\eta}{\lambda}\right);$$

hence compressing an aperture distribution, as measured in wavelengths, by factors p and q expands the associated field radiation pattern by the same factors. The similarity theorem underlies less precise rules of the form "beamwidth is inversely proportional to aperture size, and directly proportional to wavelength".

29. Shift theorem. By a different change of variable

$$P(\vartheta - \Theta, \varphi - \Phi) = \iint \left[e^{-i\,2\pi\left(\Theta\frac{\xi}{\lambda} + \Phi\frac{\eta}{\lambda}\right)} F\left(\frac{\xi}{\lambda}, \frac{\eta}{\lambda}\right)\right] e^{i\,2\pi\left(\vartheta\frac{\xi}{\lambda} + \varphi\frac{\eta}{\lambda}\right)} d\left(\frac{\xi}{\lambda}\right) d\left(\frac{\eta}{\lambda}\right).$$

Hence if an aperture distribution is modified by causing its phase to change linearly with distance across the aperture then the radiation pattern is shifted in the direction of lagging phase by an amount proportional to the phase gradient. A beam shift Θ requires a phase gradient of $2\pi\,\Theta$ radians per wavelength.

30. Array theorem. Let the convolution of the function $F(\xi/\lambda, \eta/\lambda)$ with some other function $L(\xi/\lambda, \eta/\lambda)$ be written

$$L * F \equiv \iint L\left(\frac{\xi}{\lambda} - \mu, \frac{\eta}{\lambda} - \nu\right) F(\mu, \nu)\,d\mu\,d\nu$$

---

[1] I. N. Sneddon: in vol. II of this Encyclopedia (Mathematical Methods), p. 198. Berlin 1955.
[2] H. G. Booker and P. C. Clemmow: Proc. Inst. Electr. Engrs. III **97**, 11 (1950).

and let the two dimensional Fourier transform of L be $A(\vartheta, \varphi)$, i.e.

$$A(\vartheta, \varphi) = \iint L\left(\frac{\xi}{\lambda}, \frac{\eta}{\lambda}\right) e^{-i2\pi\left(\vartheta\frac{\xi}{\lambda} + \varphi\frac{\eta}{\lambda}\right)} d\left(\frac{\xi}{\lambda}\right) d\left(\frac{\eta}{\lambda}\right).$$

Then according to the two dimensional convolution theorem

$$A(\vartheta, \varphi)\, P(\vartheta, \varphi) = \iint (L * F) \, e^{-i2\pi\left(\vartheta\frac{\xi}{\lambda} + \varphi\frac{\eta}{\lambda}\right)} d\left(\frac{\xi}{\lambda}\right) d\left(\frac{\eta}{\lambda}\right).$$

One way of interpreting this theorem is as follows. Consider an array of elements each of which has aperture distribution F and pattern P, and let the elements be at the points $(\xi_i/\lambda, \eta_i/\lambda)$. The resultant aperture distribution can be expressed as the convolution $L * F$, where

$$L = \sum_i {}^2\delta\left(\frac{\xi - \xi_i}{\lambda}, \frac{\eta - \eta_i}{\lambda}\right),$$

and ${}^2\delta(\xi, \eta)$ is the two-dimensional impulse function with the properties $\iint {}^2\delta(\xi, \eta)\, d\xi\, d\eta = 1$ and ${}^2\delta(\xi, \eta) = 0$ where $\xi^2 + \eta^2 \neq 0$. Consequently the resultant field radiation pattern of an array of elements can be expressed as the product of the element pattern P and an array factor A which is the transform of the array arrangement L.

31. Directivity. In an earlier section the directivity was defined in terms of the field radiation pattern; we now deduce its dependence on the aperture field distribution. Consider a highly directional aerial consisting of a large but finite plane aperture across which is maintained a tangential electric field E. The field radiation pattern P is the Fourier transform of E; the radiation pattern $P P^*$ is the transform of $E * E^*(-)$, by the convolution theorem. The power radiated per unit solid angle in the direction $l = m = 0$ is $P P^*|_0$ and the total power radiated is, to a good approximation, $\iint P P^* \, dl\, dm$. Hence the directivity is given by

$$D = \frac{P P^*|_0}{\dfrac{1}{4\pi} \iint P P^* \, dl\, dm}.$$

Now the infinite integral of a function is the value of its Fourier transform at the origin; hence

$$D = \frac{4\pi \iint [E * E^*(-)] \, d\left(\frac{\xi}{\lambda}\right) d\left(\frac{\eta}{\lambda}\right)}{E * E^*(-)|_0}.$$

Since the infinite integral of the convolution of two functions is the product of their infinite integrals,

$$D = \frac{4\pi \iint E \, d\left(\frac{\xi}{\lambda}\right) d\left(\frac{\eta}{\lambda}\right) \iint E^* \, d\left(\frac{\xi}{\lambda}\right) d\left(\frac{\eta}{\lambda}\right)}{\iint E E^* \, d\left(\frac{\xi}{\lambda}\right) d\left(\frac{\eta}{\lambda}\right)}$$

or, in terms of the mean value E_m over the area $\mathfrak{A}$ of the aperture,

$$D = \frac{4\pi\, \mathfrak{A}/\lambda^2}{\dfrac{\lambda^2}{\mathfrak{A}} \iint \left(\frac{E}{E_m}\right)\left(\frac{E}{E_m}\right)^* d\left(\frac{\xi}{\lambda}\right) d\left(\frac{\eta}{\lambda}\right)}.$$

In taking the total power radiated to be $\int\limits_{-\infty}^{\infty} \int\limits_{-\infty}^{\infty} P P^* \, dl\, dm$ we have assumed the integrand to fall to negligible proportions while l and m are still small com-

5*

pared with unity. If this is not true then $\int\limits_{-\infty}^{\infty}\int\limits_{-\infty}^{\infty}$ should be replaced by $\iint\limits_{l^2+m^2<1}$ since the evanescent fields for which l^2+m^2 exceeds unity do not carry away radiation. These fields are associated with the Fourier components of E which have high spatial frequencies exceeding one cycle per free space wavelength and so, if necessary, the symbol E in the preceding formula may be interpreted as that part of the field distribution not containing high spatial frequencies. We do this in a later section which treats the effect of bolt heads and other irregularities.

32. Directivity factor. If the aperture is uniformly excited all over, E/E_m is unity and $D = 4\pi\mathfrak{A}/\lambda^2$. Since we have proved that $g = 4\pi A/\lambda^2$, it follows that the effective area A of a uniform loss-free aperture is equal to its physical area $\mathfrak{A}$. When the excitation over an aperture is not uniform, the directivity falls by a factor $\mathfrak{D}$ referred to as the directivity factor (sometimes gain factor); thus

$$\mathfrak{D} = \frac{1}{\dfrac{1}{\mathfrak{A}}\iint\left(\dfrac{E}{E_m}\right)\left(\dfrac{E}{E_m}\right)^{*}d\xi\,d\eta}.$$

c) Types of aerial.

A wide variety of aerials find application in radio astronomy. In this section we review many of the types briefly with comments on their behavior and considerations entering into their design.

33. Dipoles. As used in connection with aerials, the term dipole refers to a conductor about half a wavelength long, broken in the centre, as in Fig. 19(a).

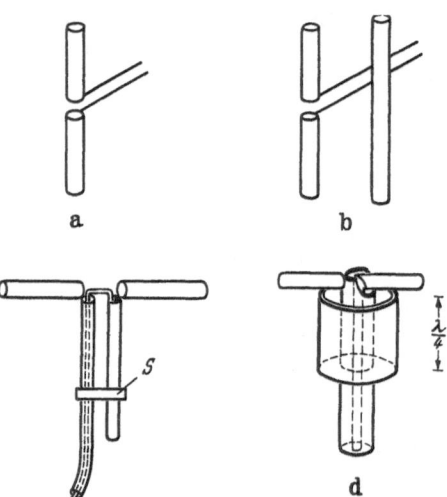

Infinitesimal dipoles are not ordinarily used. Power radiated from a dipole is concentrated towards the equatorial plane, equally in all longitudes. The radiation pattern takes the form

$$\left[\frac{\cos\left(\tfrac{1}{2}\pi\cos\alpha\right)}{\cos\alpha}\right]^2,$$

which differs only slightly from $\sin^2\alpha$, and the directivity in the equatorial plane is 1.64. As the operating frequency is changed the impedance of a dipole changes from its nominal value of about 70 ohms. However the need for tuning adjustments can be avoided over at least a two to one frequency range by fattening the conductors.

Dipoles are combined with other elements in many ways. The directivity may be doubled by placing a reflector behind the dipole in such a way that the reflected field reinforces the primary field in the forward direction. Fig. 19(b) illustrates one arrangement, but the mechanical embodiments of the idea are very numerous[1].

Fig. 19 a—d. (a) A dipole, (b) a dipole with reflector, (c) and (d) baluns.

If it is necessary to connect a coaxial cable to a dipole a balance-to-unbalance transformer or balun is incorporated to prevent unbalance currents received on

---

[1] For an extensive treatment see G. H. Brown: Proc. Inst. Radio Engrs. **25**, 78 (1937).

the outside of the cable from contributing to the signal sent down the inside. The Pawsey stub [Fig. 19(c)] achieves this result by symmetry, at the same time providing an impedance matching adjustment by means of a shorting strap S, and the quarter-wave trap [Fig. 19(d)] inserts a high impedance in the path of unbalance currents.

34. Yagi aerials. A very convenient mechanical arrangement for increasing the gain of a dipole is afforded by the arrangement of Fig. 20, which was introduced by YAGI[1]. In addition to the reflecting element there are a number of "directors" so spaced and adjusted in length that the phase of the currents induced in them causes reinforcement of radiation in the direction of the arrow.

With care, the gain of a Yagi aerial containing n elements can be made about n times that of one dipole, the adjustment being carried out by trial. Since the lengths of the elements and their spacing depend on frequency, aerials of this type are essentially limited to a narrow band of frequencies.

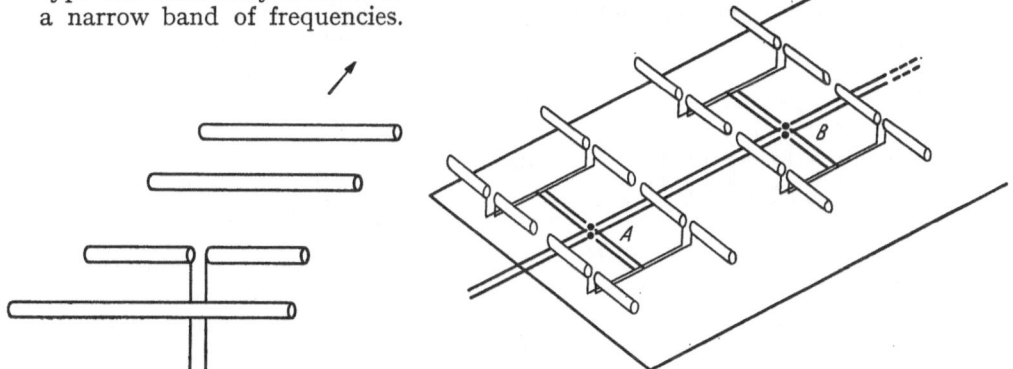

Fig. 20. A Yagi aerial.   Fig. 21. A broadside array of dipoles.

35. Broadside arrays. Arrays of dipoles such as in Fig. 21 are effective at meter wavelengths. The gain obtainable exceeds that of a single dipole by a factor roughly equal to twice the number of elements if, as is usual, a plane reflector backs up the array. The gain g per dipole-with-reflector is approximately 3, and the effective area per dipole, $g\lambda^2/4\pi$, is approximately $\frac{1}{4}\lambda^2$. This consideration illuminates the practice of spacing the dipoles at intervals of one half wavelength in both directions, but mechanical and electrical factors usually control this choice. If necessary the same gain can be obtained with substantially fewer elements. Feeding the dipoles all in equal amplitude and phase can be done by two essentially different methods both of which are illustrated in the figure. The group of dipoles fed from A are connected identically and therefore remain equally excited when the frequency is changed. But correct excitation of the group connected to B will require a certain relation between the length AB and the wavelength. Such frequency-dependent arrangements are often adopted for simplicity.

The effective area of a large broadside array can be made very nearly equal to the physical area, and the possibility of approaching the theoretical design in practice is an advantage where absolute measurements of radiation intensity are attempted.

Arrays of large size can be pointed in different azimuths but equatorial mounting, with few exceptions has not generally been deemed feasible. However

[1] H. YAGI: Proc. Inst. Radio Engrs. **16**, 715 (1928).

electrical beam-swinging is readily incorporated. Suppose for example that the section $A B$ in Fig. 21 had a different electrical length, i.e. the phase difference $(A B) \times (2\pi/\lambda)$ between A and B were different; then, as explained above in connection with the shift theorem, the beam would be shifted in proportion. Beamswinging is accomplished by actual physical adjustments or by changing wavelength.

36. Helical aerials. The aerial illustrated in Fig. 22 is akin to the Yagi, in that it is extended in the direction of radiation (indicated by the arrow), but has important differences which favor its application in radio astronomy. It is not as sensitive to changes in frequency, ranges of nearly two to one in frequency being not unreasonable. Careful adjustment is not required, a coaxial cable may be used without a balun, and high directivity is achieved with mechanical simplicity. The ground plane disc may be less than a wavelength in diameter.

Under correct operating conditions a wave travels around the helical wire with such a phase velocity that the phase difference between corresponding points on adjacent turns exceeds $2\pi d/\lambda$ by one cycle, d being the pitch of the helix. The distant fields of successive turns then reinforce in the axial direction and produce circular polarization.

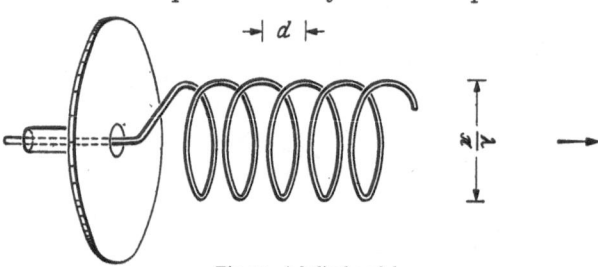

Fig. 22. A helical aerial.

From a combination of theory and measurement Kraus[1] gives the half power beamwidth in degrees as $52(\lambda/C)(\lambda/L)^{\frac{1}{2}}$, where C is the circumference and L the length of the helix, provided the circumference is about one wavelength, the pitch about one quarter wavelength, and the number of turns greater than 3. From this we can say that the directivity will be approximately $15(C/\lambda)^2(L/\lambda)$ and the effective area $15(L/\lambda) \times$ (projected area normal to axis). The input impedance is about 140 ohms.

37. Horns. The open end of a rectangular waveguide radiates very much as one would expect from a plane aperture on which is maintained an electric field distributed as indicated in Fig. 23 (a). When the end of the waveguide is flared as in Fig. 23 (b) greater directivity results, and provided the aperture dimensions are less than about 3 wavelengths the electric field in the aperture can be written

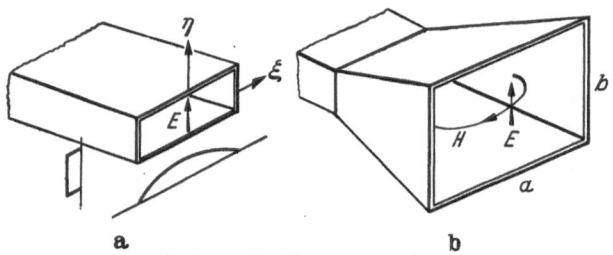

Fig. 23 a and b. Electromagnetic horns.

$$E_\eta = \frac{\pi}{2} \cos \frac{\pi \xi}{a} \, \Pi\left(\frac{\xi}{a}\right) \Pi\left(\frac{\eta}{b}\right),$$

where $\Pi(\xi)$ is the rectangle function of unit height and width, i.e.

$$\Pi(\xi) = \begin{cases} 1 & (|\xi| < \tfrac{1}{2}), \\ 0 & (|\xi| > \tfrac{1}{2}). \end{cases}$$

---

[1] For details and further references see J.D. Kraus: Antennas. New York 1950.

From the Fourier transform relation we have shown that the directivity is reduced, relative to a uniformly excited aperture for which

$$E_\eta = \Pi\left(\frac{\xi}{a}\right)\Pi\left(\frac{\eta}{b}\right).$$

by a factor $\mathfrak{D}$. Having normalized E_η to have unit mean value and to cover unit interval, we find succinctly,

$$\mathfrak{D} = \left[\int_{-\frac{1}{2}}^{\frac{1}{2}}\left(\frac{\pi}{2}\cos\pi\xi\right)^2 d\xi\right]^{-1} = \frac{8}{\pi^2}.$$

Hence the effective area is $8ab/\pi^2$, or about 0.8 of the physical area. The beamwidth in degrees between points where the radiation pattern is one tenth of the maximum is given by

$$\frac{88\lambda}{b} \; (\eta\zeta \text{ plane}),$$

$$31 + \frac{79\lambda}{a} \; (\xi\zeta \text{ plane}).$$

The value of these empirical formulae (SILVER[1]) will be seen below in connection with feeds for parabolic reflectors.

When the aperture of a horn is larger than about 3 wavelengths, there will be a phase inequality across the aperture depending on the length of the flared section. The directivity reduction integral corresponding to that given above can be evaluated and SCHELKUNOFF and FRIIS[2] present convenient charts for determining the directivity.

One of the few applications in radio astronomy of a horn alone was to the discovery of the galactic hydrogen line emission by EWEN and PURCELL. Usually a large horn requires too much volume for a given effective area. There are, however, some important uses. One is the provision of a calculable standard of gain intermediate between that of a dipole and that of an aerial under calibration. Another is for feeding paraboloidal reflectors. The high efficiency and relative freedom from wide-angle side radiation also suit them for use with future noise-free receivers.

38. Paraboloidal reflectors. By placing a source at the focus of a reflecting paraboloid of revolution one may produce extensive equiphase fields in planes normal to the axis. In comparison with a broadside array the means used are simple; there is only one radiator and one feeder. Furthermore, the frequency of operation can be altered by readjustment or replacement of the single feed system. Thus for many purposes a paraboloidal reflector offers advantages.

At the focus one commonly finds a dipole and reflector, or a horn; however any aerial may be used including small broadside arrays in the larger reflectors at longer wavelengths. The effective solid angle Ω_f of the feed antenna must harmonize with the solid angle subtended by the reflector at its focus. If Ω_f is too small, the outer part of the reflector is not fully utilized, and directivity is lost; while if Ω_f is too large there is "spillover", which has two effects. First there is a pure loss of energy in undesired directions and then, because of the large discontinuity in illumination at the edge of the reflector, the side radiation, though reduced, takes on a pronounced lobe structure. It is therefore customary to taper the illumination towards the rim, balancing the loss of directivity

S. SILVER: Microwave antenna theory and design. New York 1949.
S. A. SCHELKUNOFF and H. T. FRIIS: Antennas, theory and practice. New York 1952.

against the undesirability of distinct sidelobes. Equating the angle subtended at the focus by the rim of the paraboloid to the 10-decibel beamwidth of a feed horn, we have

$$4 \operatorname{arc cot} \frac{4F}{D} = 88 \frac{\lambda}{b} = 31 + 79 \frac{\lambda}{a},$$

where F/D is the ratio of focal length to diameter of the paraboloid. These equations give suitable dimensions, a and b, for the feed horn. However, other proportions hould be considered for partic-

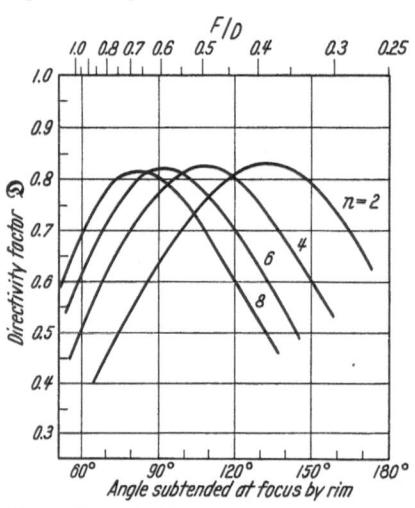

ular purposes, and some of the factors governing the design follow.

Whether the loss in directivity factor is quantitatively serious may be determined from Fig. 24 (adapted from Silver), which gives the theoretical directivity factor for the normalized set of calculable feed patterns $2(n+1) \Pi(\vartheta/\pi) \cos^n \vartheta$. One of these functions will represent the actual feed pattern over most of the beam and Fig. 24 then gives the optimum focal ratio, and an approximate value of $\mathfrak{D}$. If desired, a better value of $\mathfrak{D}$ can then be calculated from an improved determination of the aperture field distribution which takes into account the actual feed pattern and allows for the inverse squares of the path lengths from the focus to the paraboloid.

Fig. 24. Theoretical directivity factor as a function of focal ratio F/D for different values of the taper parameter n in the feed pattern $2(n+1) \Pi(\vartheta/\pi) \cos^n \vartheta$.

The aerial feeding a paraboloid will normally have been adjusted to match its transmission line in the absence of the reflector, but when it is placed in the focus of a small paraboloid and excited by a generator it intercepts some reflected energy which then causes standing waves on the line. The fraction of power

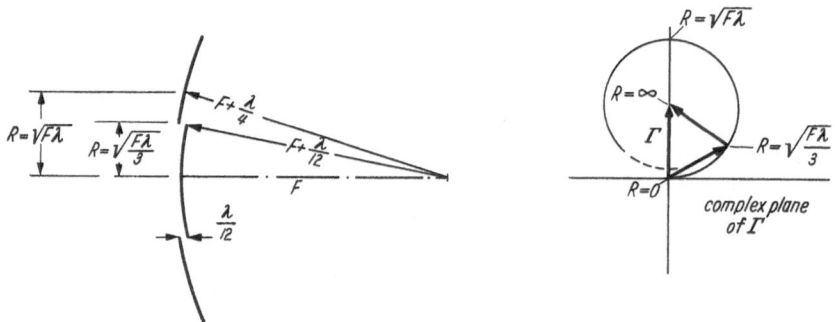

Fig. 25. Matching a paraboloid with a vertex plate.

intercepted will be approximately equal to A_H/A_P the ratio of the effective areas of the feed horn and the paraboloid and so the power reflection coefficient $|\Gamma|^2$ observed on the transmission line will approximately equal this ratio. Putting $A_P = F^2 \Omega_H$, where $4\pi/\Omega_H = g_H = 4\pi A_H/\lambda^2$, we have

$$|\Gamma| = \frac{g_H \lambda}{4\pi F} = \frac{A_H}{F\lambda},$$

alternative expressions for reflection coefficient which do not involve the aperture of the parabola. The reason for this is that the first few Fresnel zones around the vertex, which contribute the bulk of the reflection, occupy only a small fraction of the aperture. The radius of the first Fresnel zone, assuming a radius of curvature $2F$, is $\sqrt{F\lambda}$, and a cap of radius $\sqrt{F\lambda/3}$ contributes a reflection equal in magnitude to the full reflection but leading it in phase by a sixth of a cycle, as shown in Fig. 25. The region outside the cap contributes an equal amount which lags the full reflection by a sixth of a cycle. If now this cap is moved towards the focus by a twelfth of a wavelength, as in the illustration, the total reflection will be annulled. In practice a flat plate of the calculated radius is placed at the mean position of the cap, i.e. $\lambda/8$ from the vertex, any further adjustments being made by trial.

The foregoing considerations are complicated by mechanical details of the structure supporting the feed, which it may only be possible to take into account empirically. The following procedure nevertheless reveals some of the effects to be expected. Cancel the aperture distribution over the geometrical shadow cast by all the structures, including the feed horn, in front of the reflector. Then subtract from the field radiation pattern the pattern due to an aperture the same shape as the shadow.

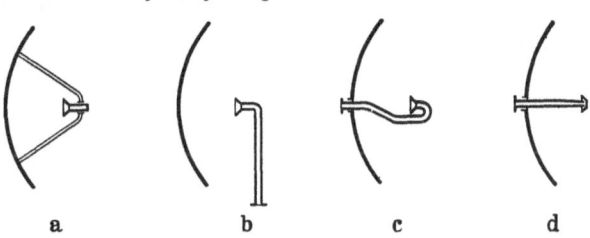

a   b   c   d

Fig. 26 a—d. Methods of supporting a feed horn at the focus.

The principal effects thus deduced are a reduction in effective area by the shadowing and an increase of side radiation in a distinctive arrangement associated with the shadow shape. The ultimate destination of the radiation impeded by the structure can also be considered, especially if it leads directly back into the horn.

Some basic arrangements are shown in Fig. 26. In (a) the horn is supported on four struts rising from firmly held points on the reflector, and the transmission line from the horn runs down one of the struts. In this very satisfactory arrangement a main concern is to minimize the shadow cast by parts of the structure near the horn. In (b) shadowing is small though asymmetrical, and the arrangement is ideal where access to the feeder near the rim is convenient. Arrangement (c) is suitable when access at the vertex is desired and it is simple mechanically. There is substantial asymmetrical shadowing and a direct reflection into the horn which must be minimized experimentally. In (d) an excellent mechanical arrangement is obtained at the cost of considerable empirical design and some restriction on bandwidth associated with the compact T-junction and pair of reflex horns at the focus.

Paraboloidal aerials receive some stray energy over and above that which is focused by reflection from the paraboloid. There is some side-reception by the collecting aerial at the focus from directions outside the cone defined by the rim of the reflector, and there may be some transmission through the reflecting surface itself (shine-through) if it is not a continuous metal sheet. There is also some scattering from the roughnesses of the reflector, and from the feed support and other incidental structures.

Let power W_{in} enter the terminals of an aerial such as is shown in Fig. 26. Some is absorbed in the transmission line; of that which emerges from the feed horn let W_{beam} be the amount intercepted by the paraboloid and focused into the main beam. We use the term main beam to refer to the radiation launched, ideally, with the characteristic Fraunhofer diffraction pattern of the circular aperture. Let W_{sky} be the total power launched towards the sky whether by the main beam, or by stray routes such as direct or ground-reflected spillover, scatter, or shine-through. Then the beam ratio $\mathfrak{B}$ is defined by

$$\mathfrak{B} = \frac{W_{\mathrm{beam}}}{W_{\mathrm{sky}}}$$

and, by the definition of efficiency η in Sect. 25,

$$\eta = \frac{W_{\mathrm{sky}}}{W_{\mathrm{in}}}.$$

We now calculate the aperture efficiency α, the ratio of the effective area A to the area $\mathfrak{A}$ of the circle bounded by the rim of the paraboloid. From Sect. 25 and 26, the gain g is given by

$$\frac{4\pi A}{\lambda^2} = \frac{W(0,0)}{\frac{1}{4\pi}W_{\mathrm{in}}} = \frac{W(0,0)}{\frac{1}{4\pi}W_{\mathrm{beam}}}\frac{W_{\mathrm{beam}}}{W_{\mathrm{sky}}}\frac{W_{\mathrm{sky}}}{W_{\mathrm{in}}} = \mathfrak{D}\,\frac{4\pi\mathfrak{A}}{\lambda^2}\,\mathfrak{B}\eta,$$

where the directivity factor $\mathfrak{D}$, referred to the rim of the paraboloid, is given as in Sect. 32 by

$$\mathfrak{D} = \frac{1}{\frac{1}{\mathfrak{A}}\iint\left(\frac{E}{E_m}\right)\left(\frac{E}{E_m}\right)^{*}d\zeta\,d\eta}.$$

(In evaluating this expression small-scale irregularities in E that scatter energy out of the main beam should be ignored.) Hence for the aperture efficiency α we have

$$\alpha = \frac{A}{\mathfrak{A}} = \mathfrak{D}\,\mathfrak{B}\,\eta.$$

The factor $\mathfrak{B}\eta$ is the beam efficiency β, the ratio of the power launched in the beam to that entering the aerial terminals:

$$\beta = \mathfrak{B}\eta = \frac{W_{\mathrm{beam}}}{W_{\mathrm{in}}}.$$

The application of these parameters is discussed in Sect. 69 and 70. Their measurement involves one step beyond the determination of g and η, such as, for example, a measurement of the shape of the main beam. Thus the presence of stray radiation will cause the effective solid angle Ω_{beam} of the main beam to be less than the expected value $4\pi/D = \lambda^2/\mathfrak{D}\,\mathfrak{A} = 4\pi\eta/g$. Hence

$$\mathfrak{B} = \frac{\mathfrak{D}\,\mathfrak{A}\,\Omega_{\mathrm{beam}}}{\lambda^2}$$

and

$$\beta = \frac{g\,\Omega_{\mathrm{beam}}}{4\pi}.$$

The effective solid angle of the main beam can be determined from careful non-absolute pattern measurement over a limited region. Then Ω_{beam} is the double integral of the limited reception pattern divided by the axial value. The integral is extended over a region sufficiently large, usually a few beamwidths in extent, that further extension makes an inappreciable difference for the purpose in hand.

Fig. 24 shows that $\mathfrak{D}$ will commonly be around 70 per cent. Beam ratios $\mathfrak{B}$ around 60 per cent are common and as the operating frequency is raised the surface irregularities of a given reflector introduce a factor ζ (Fig. 28) that leads to even smaller values. An efficiency η of 10 to 50 per cent is commonly accepted but much improved values are obtainable by eliminating the transmission line between the horn and the receiver. Further improvement in efficiency is obtainable by enlarging the horn to reduce spillover absorption by the ground; this reduces side-reception of thermal emission from the ground by the horn, but reduces $\mathfrak{D}$. The highest possible efficiency η is needed for sensitive observations in order to minimize the effective receiver noise temperature $(N/\eta - 1)\,T_0$ mentioned in Sect. 15, and to minimize objectionable changes of ground radiation as the aerial moves. This premium on efficiency calls for horn designs, for use with noise-free amplifiers, that accept a reduction in aperture efficiency α in favor of optimum performance with higher η.

d) Feeders.

39. Transmission lines and waveguides. The transmission line or feeder associated with an aerial may often conveniently be considered as a part of the aerial.

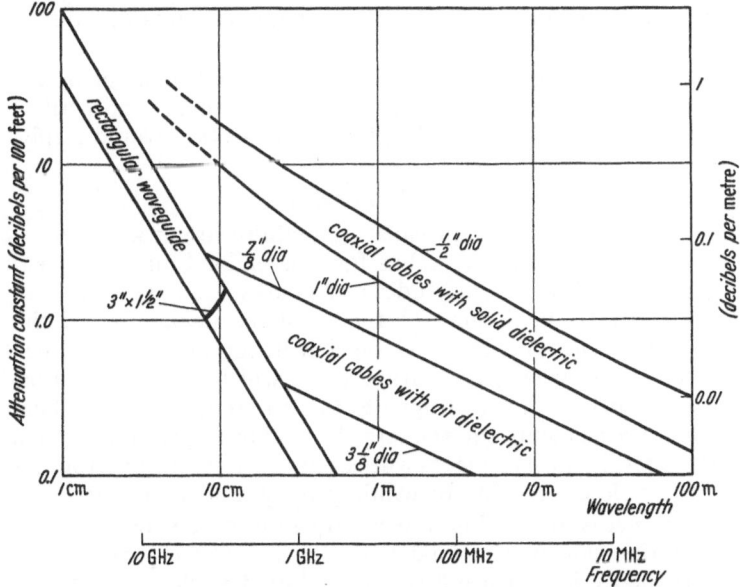

Fig. 27. Attenuation constant and range of use of various feeders.

It may serve an impedance matching function, be adapted for standing wave measurement, and enter into switching, coupling, and phasing devices.

In Fig. 27 the attenuation constant of various available feeders is shown. Flexible coaxial cables with solid dielectric are usable at all but the highest frequencies, where the dissipation in conductor and dielectric becomes prohibitive. Coaxial lines with a rigid outer conductor instead of braid, and with air dielectric, furnish lower attenuations especially in the larger diameters. However, when the mean circumference of the air space exceeds the wavelength the appearance of a second transmission mode renders the line unusable. There is therefore an upper limit to both diameter and frequency. Beyond the useful range of coaxial lines, and in a transition region, rectangular waveguide may be used. Unlike the coaxial line, which is satisfactory over a semi-infinite spectrum, each waveguide size is restricted to less than one octave. The figure shows a curve for the waveguide used at 10 cm; similar curves for the waveguide appropriate to each wavelength fall within the zone indicated.

Some other lines, not shown in Fig. 27, are important. Two wire transmission line has about the same conductor losses as coaxial line of similar dimensions and is far better mechanically in many ways. It cannot be used where complete shielding is essential, is subject to radiation loss at the highest frequencies, and is affected seriously by dew and other natural phenomena. It has been used successfully in very long runs for experimental purposes as high as 1400 MHz.

Ridged waveguide[1] permits operation over a range of two octaves, strip line[2] offers many convenient features, and single-wire surface wave transmission line[3] offers the possibility of extremely long straight runs for which waveguide would be too costly.

An important consideration in radio astronomy is stability of the phase length of a feeder, a quantity which may be affected by humidity, temperature, mechanical effects, and frequency change. Measures to achieve stability include sealing off from the air, pressurizing, burying or shading, heating, tensioning, and broadbanding connectors, bends, junctions, supports and other discontinuities in the line.

A large fraction of the essential technique in the development of a successful instrument for radio astronomy is spent on the design and adjustment of the transmission system, more than the length of this section would imply.

e) Tolerances.

Consider an aerial whose design calls for definite amplitudes and phases in a certain plane. When the structure has been assembled and ajusted there will be departures from the desired amplitudes and phases due to the following causes. In the case of a paraboloidal reflector there will be departures of the surface from the true paraboloid and the feed horn will not be perfectly located and oriented. In the case of a broadside array there will be small reflections on the feeders and discrepancies in phase length to the different elements. The location and orientation of the elements and the reflector will be imperfect, and in addition, the sizes of all the mechanical elements will be subject to variation from the design. There will be variable effects due to wind, gravity, and temperature.

It is therefore necessary to know what tolerances may be allowed in order to limit the deterioration in the final behavior to an acceptable degree. This branch of aerial theory is in a primitive state, but it is clear that the large and complex aerials of radio astronomy will benefit from a suitable theory of tolerances. It has been customary to measure the gain and side radiation of actual aerials and to make empirical adjustments, but the adjustments become more laborious with increasing aerial size and the measurements themselves are becoming infeasible.

In the following discussion it will be assumed that the amplitude and phase errors are known and the effect on the radiation pattern, and especially the directivity and beamwidth, will be studied.

---

[1] Very-high-frequency Techniques. New York-Toronto-London: MacGraw-Hill 1947.
[2] See various papers in Transactions of the IRE Professional Group on Microwave Theory and Techniques, vol. MTT 3, March 1955.
[3] G. Goubau: Electronics **27**, 180 (1954).

40. The directivity possible of achievement. It was shown earlier that the directivity factor $\mathfrak{D}$ of an aperture of physical area $\mathfrak{A}$ is given by

$$\mathfrak{D} = \frac{\mathfrak{A}}{\iint \left(\frac{E}{E_m}\right)\left(\frac{E}{E_m}\right)^* d\xi\, d\eta}\,.$$

It is convenient to express the aperture illumination E relative to the mean value E_m over the aperture because the type of maladjustment which would cause the whole illumination to be say ten per cent low or ten degrees lagging in phase does not affect the directivity. Therefore, by first normalizing with respect to the mean we can concentrate separately on the quantity of interest.

Consider the case where the departures from the mean illumination E_m are εE_m. Then

$$\mathfrak{D} = \frac{\mathfrak{A}}{\iint (1 + \varepsilon)(1 + \varepsilon)^* d\xi\, d\eta}$$

$$= \frac{1}{1 + \dfrac{1}{\mathfrak{A}} \iint \varepsilon\, \varepsilon^* d\xi\, d\eta}$$

$$= \frac{1}{1 + \operatorname{var}\varepsilon}\,,$$

where $\operatorname{var}\varepsilon$ is the variance, or mean square modulus, of the complex fractional departures ε. This basic formula, which is valid for both systematic and random departures in both phase and amplitude, has valuable applications.

If a non-uniformly illuminated aperture is perturbed so that the fractional departures from the perturbed mean are ε' and the perturbed directivity factor is $\mathfrak{D}'$, then the factor ζ by which the perturbation reduces the directivity is given by

$$\zeta = \frac{\mathfrak{D}'}{\mathfrak{D}} = \frac{1 + \operatorname{var}\varepsilon}{1 + \operatorname{var}\varepsilon'}\,.$$

Conversely defined, ζ is the factor which measures the extent to which $\mathfrak{D}'$ achieves the design value $\mathfrak{D}$; thus

$$\mathfrak{D}' = \zeta\mathfrak{D}.$$

We shall refer to ζ as the directivity achievement factor. In terms of a minimum acceptable value, such as 0.9, we can determine tolerances which are permissible.

Let us consider first the case of an aperture designed to be uniformly illuminated with a mean field E_m, but which is subject to an undesired phase error δ which varies from point to point over the aperture in any way. This is similar to the situation described in optics by a wavefront which departs from the ideal sphere, and it engenders all the types of aberration familiar in optical systems (MARÉCHAL[1]). The perturbed illumination $E_m \exp i\delta$ has a lowered mean value $E'_m \approx E_m (1 - \frac{1}{2} \operatorname{var}\delta')$, for small δ', where δ' is referred to the new mean phase, should it be altered, i.e.

$$\delta' = \delta - \frac{1}{\mathfrak{A}} \iint (1 + \varepsilon)\, \delta\, d\xi\, d\eta;$$

but the mean square modulus of the perturbed illumination is unchanged. Therefore

$$\zeta = \left(\frac{E'_m}{E_m}\right)\left(\frac{E'_m}{E_m}\right)^* \approx 1 - \operatorname{var}\delta'.$$

---

[1] A. MARÉCHAL: in vol. XXIV of this Encyclopedia (Fundamentals of Optics), p. 44. Berlin 1956.

Alternatively we may say that the change in absolute mean value is small and that $E'_m(1+\varepsilon') = E_m \exp i\delta \approx E'_m(1+i\delta')$, whence $\varepsilon \approx i\delta'$. Then, with $\varepsilon \approx 0$

$$\zeta = \frac{1 + \text{var } \varepsilon}{1 + \text{var } \varepsilon'} \approx \frac{1}{1 + \text{var } \delta'}.$$

Both lines of reasoning are useful for obtaining approximations in non-uniform cases; however, we can state without approximation that an aperture which is intended to be uniformly illuminated and which in fact has fractional departures ε' from the mean, of any magnitude, whether in amplitude or phase, has

$$\zeta = \frac{1}{1 + \text{var } \varepsilon'}.$$

In a non-uniform case it may be possible to calculate the directivity factor in both the perturbed and unperturbed situations. If not, various approximate formulae may be evolved; for example if there are small phase errors δ, then by expanding the exponential factor in

$$E'_m = \frac{1}{\mathfrak{A}} \iint E\, e^{i\delta}\, d\xi\, d\eta$$

we find

$$\zeta \approx 1 - \text{var } \sqrt{1 + \varepsilon}\, \delta'.$$

41. Squint. Let there be a progressive phase error across the aperture of amount α radians per meter in the x direction. If the extreme phase error is small, we have by the preceding equation, a reduction in directivity given by

$$\zeta \approx 1 - \alpha^2 \text{var } \sqrt{1 + \varepsilon}\, x.$$

More precisely

$$\zeta = \left|\frac{E'_m}{E_m}\right|^2 = \left|\frac{1}{\mathfrak{A}} \iint (1 + \varepsilon) \exp i\alpha\xi\, d\xi\, d\eta\right|^2$$

$$= \frac{PP^*|_\Theta}{PP^*|_0}$$

where $\Theta = \alpha\lambda/2\pi$; i.e. the whole reduction is due to a displacement of the radiation pattern PP^* through an angle $\alpha\lambda/2\pi$ without change of shape. The angle of squint Θ is the same as that which would have been predicted by the shift theorem.

42. Quadratic phase error. Let $\delta = \beta\xi^2$. Then different effects will result according as the aperture is uniformly illuminated or not. In the case of a uniformly illuminated rectangular aperture the mean phase of the perturbed illumination is one third of the phase error $\varDelta$ at the edge of the aperture: Hence

$$\zeta = 1 - \text{var } \delta'$$

$$= 1 - \text{var}\left(\beta\xi^2 - \frac{\varDelta}{3}\right)$$

$$= 1 - \frac{4\varDelta^2}{45};$$

and for $\zeta = 0.9$ evidently an extreme phase error of 1.1 radians can be tolerated in the uniform case and even more if the illumination is tapered.

Now let $\delta = \beta(\xi^2 + \eta^2)$ over a uniformly illuminated circular aperture. Then the mean phase is one half the phase error $\varDelta$ at the edge and

$$\zeta = 1 - \operatorname{var}\left[\beta(\xi^2 + \eta^2) - \frac{\varDelta}{2}\right]$$
$$= 1 - \frac{\varDelta^2}{12},$$

which is approximately the same as before.

Non-uniform illumination may be handled as indicated already or treated in a way mentioned below.

43. Random errors. Since there is nothing in the preceding work which excludes random errors the formulae given may be adopted immediately. In fact, the expressions in terms of mean square departures from a mean are especially appropriate. For example, a uniform broadside array whose elements are subject to small phase errors δ' from the mean has an achievement factor $1 - \operatorname{var}\delta'$; thus for $\zeta = 0.9$ a root-mean-square phase error of 0.3 radians (root-mean-square path difference of $\lambda/30$) can be tolerated. If the elements are in phase but unequally excited, then a root-mean-square fractional error of 0.3 can be tolerated; and if both sorts of error are present then the tolerances are 0.2 radians ($\lambda/28$) and $\pm 20\%$. These tolerances are liberal, especially when it is considered that extreme errors, both positive and negative, may substantially exceed the root-mean-square value. Whilst the illustrations given here are taken from convenient approximate formulae, it is recalled that, whether the errors are large or small, recourse may be had to the accurate result $\zeta = (1 + \operatorname{var}\varepsilon')^{-1}$. This result, moreover, does not assume a normal distribution of the errors, or any particular spatial correlation. There is, however, an important point of interpretation in the case of continuous reflectors.

As pointed out in the section on directivity, evanescent fields do not normally have importance in the angular spectrum of a highly directional aerial, but if they do, the symbol E in the formula for directivity may be interpreted as the part of the field distribution remaining when Fourier components of spatial wavelength less than the free-space radiation wavelength have been filtered out. Now in a reflector having irregularities such as bolt heads, ribs, or dents, evanescent fields are set up which should be ignored in evaluations of the distant field. Those spatial components of the irregularities whose scale is finer than the wavelength should therefore be rejected before calculating ζ; this conclusion rests on the assumption that an irregularity merely introduces a corrugation of the same shape in the wavefront, which seems appropriate to this first order discussion of smaller effects. Starting from a given corrugated wavefront one would first filter out the fine detail numerically, then evaluate $\operatorname{var}\delta'$. Alternatively, one might start from the (unnormalized) autocorrelation function[1] $\delta' \star \delta'$ of the phase corrugations referred to their mean. Then $\delta' \star \delta'|_0$, which is proportional to $\operatorname{var}\delta'$, is equal to $\iint F \, d\mu \, d\nu$, the infinite integral of F, the Fourier transform of the autocorrelation function of δ'. Then the variance of the filtered part of δ' will be proportional to the integral of that part of F for which $\sqrt{\mu^2 + \nu^2}$, the number of cycles per unit distance of wavefront, does not exceed λ^{-1}. An expression of

---

[1] We use the unobtrusive notation

$$f \star g = \iint f(u - \xi, v - \eta)\, g(u, v)\, du\, dv$$

for the unnormalized cross-correlation function of f and g as distinct from the convolution integral

$$f * g = \iint f(\xi - u, \eta - v)\, g(u, v)\, du\, dv.$$

this theoretical procedure, in terms of the rectangle function Π (Sect. 37), is

$$\zeta \approx 1 - \frac{\iint \Pi\left(\tfrac{1}{2}\lambda\sqrt{\mu^2 + \nu^2}\right) F \, d\mu \, d\nu}{\iint F \, d\mu \, d\nu} \text{ var } \delta'.$$

It may be applied to the problem of the variation of ζ of a uniform reflector as λ diminishes. Let the departures from the mean be h, then var $\delta' = (4\pi/\lambda)^2$ var h and for frequencies high enough to perceive the irregularities in full ζ will fall off parabolically as in the broken curve of Fig. 28. At low frequencies the exact form of $h \star h$ will determine the extent to which var δ' is diminished but in general a result such as the heavy curve indicates would be obtained. The point P marks the transition from mainly unresolved irregularities to nearly complete resolution.

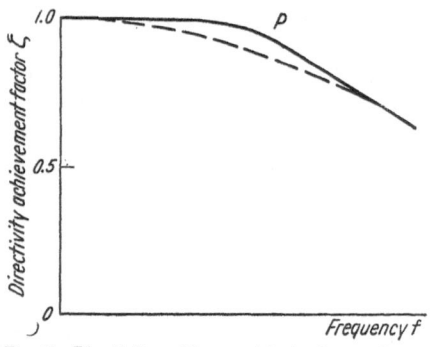

Fig. 28. Directivity achievement factor for a reflector. The broken curve corresponds to complete resolution of the irregularities.

The effect of random errors on aerials has not been widely known. Ruze[1] has made an elegant experimental study and clearly explained the principal effects. He quotes the approximate expression $1 - \text{var } \delta'$, attributing it to R. C. Spencer, and examines an extension of the formula to a particular case of partial resolution of random irregularities. When an aerial has been designed for fractional departures ε from the mean uniform illumination and the departures are in fact $\overline{\varepsilon + \varepsilon_r}$, where the ε_r are randomly distributed about zero mean value, then $\overline{\text{var}(\varepsilon + \varepsilon_r)} = \text{var } \varepsilon + \overline{\text{var } \varepsilon_r}$, where bars indicate the average over a batch, provided that ε_r is not correlated with ε on the average over a batch. Then we find the following general result:

$$\bar{\zeta} = \frac{1 + \text{var } \varepsilon}{1 + \text{var } \varepsilon + \overline{\text{var } \varepsilon_r}}$$

$$= \frac{1}{1 + \mathfrak{D}\, \overline{\text{var } \varepsilon_r}}.$$

When $\varepsilon_r \approx \dfrac{i\,4\pi h}{\lambda}$,

$$\bar{\zeta} = \frac{1}{1 + \left(\dfrac{4\pi}{\lambda}\right)^2 \mathfrak{D}\, \overline{\text{var } h}}.$$

In applying these formulae, and those below, ε_r or h must first be filtered.

It is not only the effect on ζ which is important in a study of tolerances; we are also concerned with the way in which the radiation pattern deteriorates, i.e. what happens to the shape of the main beam and what happens to the level of side radiation. A decrease in directivity implies by definition an increase in effective solid angle $\Omega = 4\pi/D$, and this may be interpreted as a widening of the beamwidth to half power in the case of errors of very low spatial frequency. However, by resolving the errors into spatial Fourier components, as described by the function F, we shall see precisely how much power is scattered into each off-axis direction. Should this scattered power be comparable with the intended radiation in a given direction; then the radiation pattern will be complicated. But if the scattered radiation predominates then we have an important property

[1] J. Ruze: Nuovo Cim. 9, Suppl. No. 3, 364, (1952).

of the undesired side radiation level immediately in terms of P, or $\delta' \star \delta'$, the autocorrelation function of the phase departures from the mean. For irregularities containing fine detail one would expect the side radiation level not far outside the main beam to be dominated by the irregularity contribution, but the shape of the main beam would not be much affected. For systematic errors such as those discussed in following sections, the side radiation would be mainly unchanged and the main beam would be fattened.

The following interesting result is obtained when numerous small roughnesses such as rivet heads scatter side radiation equally in all directions. Then the scattered side radiation level, expressed as a fraction of the aerial radiation level, is given by

$$2\frac{\frac{1}{\zeta} - 1}{D} = \frac{2\,\mathfrak{D}\,\mathrm{var}\,\varepsilon_r}{D} = \frac{2\pi\mathfrak{D}}{A}\,\mathrm{var}\,h.$$

As examples of the application of these relations, suppose that an aerial has been designed for a maximum sidelobe level 20 decibels down on the main beam, and for a directivity $D = 100$. Then if the side radiation due to isotropically scattering roughnesses is to be kept negligible, say to one part in 1000 of the axial radiation level, then ζ must be kept up to 0.9. This is a lower limit for ζ since anisotropic scattering will mean stronger scattering than the average in some directions. If the directivity is much greater than 100, it will be seen that scattered radiation is unlikely to be important. Conversely stated, larger tolerances on roughness are permissible. The expression $\dfrac{2\pi\mathfrak{D}}{A}\,\mathrm{var}\,h$ gives this in the case of a reflector and it will be noted that, provided the irregularities are fully resolved the side radiation level does not depend on wavelength.

44. Extrafocal errors. When the feed horn of a paraboloid is displaced from the focus there are two different effects according as the displacement is transverse or axial.

A given transverse displacement causes a progressive phase error across the aperture which can be written explicitly from the geometry, and which is approximately of the form

$$\delta = \alpha\,\xi - \beta\,\xi^3.$$

The first term causes a simple shift of the radiation pattern and the second term, which can be investigated as was done above for quadratic error, causes a loss of directivity. However this loss is small, because the cubic errors are strong near the edge, where the illumination is reduced.

A good deal of transverse extrafocal displacement can thus be tolerated since in most aerial applications, if not all, the axis of the radiation pattern is located by electrical observation, and not by reference to the mechanical axis.

The effect on ζ, if desired, can readily be evaluated for a given illumination and focal ratio. However by the time ζ falls appreciably below unity the displacement is so great that complications set in.

Axial displacement introduces a phase error approximately of the form

$$\delta = \alpha\,\xi^2 - \alpha_2\,\xi^4.$$

Arguing from the quadratic term alone, which has already been studied, and neglecting illumination taper, we can say that an axial displacement of order $\lambda/3$ can be tolerated for a 10% loss in directivity.

45. Aparaboloidal aberration. It often happens that an intended paraboloid departs from the truth but remains a figure of revolution. If the departure is quadratic the shape remains paraboloidal but with a changed focal length. Therefore the effect is the same as for axial displacement of the feed from the focus and can be compensated for if the error is a permanent one such as might arise during construction, and not such as would be caused by thermal expansion or wind loading. If the departure is more general then the focus not only shifts but elongates into a line segment whose ends represent the foci of paraxial and extreme rays respectively. The best compensation would result from placing the feed between the ends of the segment as reminiscent of the position of best focus under conditions of spherical aberration in optics, a closely related phenomenon. Clearly one can set limits on allowable aparaboloidal distortions, including the astigmatic, along the approximate lines indicated here, or one can refer back to the actual geometry and apply the basic formula.

46. Side radiation. Away from the main beam the side radiation pattern exhibits numerous more or less distinct minima the regions between which are referred to as sidelobes. Unwanted natural or man-made point sources may give trouble by coming in on sidelobes, which therefore receive much attention. To calculate the side radiation one takes the Fourier transform of the aperture distribution and adds to it the direct radiation from the feed which is not intercepted by the reflector. In general no accuracy at all can be expected because small errors in assumptions regarding physical dimensions, etc., cause important effects at large path differences such as exist well away from the axis, where numerous large field components combine in a state of approximate cancellation Any method of reducing the off-axis response relative to the on-axis response essentially involves an increase in directivity. Sometimes however alleviation of sidelobes is sought by tapering the aerial excitation towards the edge which, as we have seen, reduces the directivity. Evidently this is only appropriate where obe structure of the side radiation, not its strength, is a cause of concern.

f) Mountings and drive systems.

47. Mountings. The principal mountings in use are meridian, altazimuth, and equatorial. A large meridian telescope at Pott's Hill is shown in Fig. 29, one meridian-mounted element of the giant Cambridge[1] interferometer is shown in Fig. 30 and Fig. 31 shows an extended array of meridian-mounted Yagis at Stanford.

Altazimuth mountings are illustrated in Figs. 32 to 34 which show the radio telescopes at Dwingeloo, Manchester[2], and Stanford.

For almost all astronomical purposes an equatorial mounting would be preferred, but for reasons of economy altazimuth mountings are accepted, even to the extent of embodying computing mechanisms, as has often been done, for converting coordinates. Meridian mountings are even more economical and have permitted larger aerials for a given effort in applications where the restricted motion has sufficed. Of course even larger reflectors are made possible by sacrificing all motion as in the case of the 212 foot fixed paraboloid at Manchester and the 80 foot fixed paraboloid at Sydney, both of which achieved discoveries depending on high resolution many years before the advent of steerable reflectors of comparable size.

---

[1] See M. Ryle and A. Hewish: Mem. Roy. Astronom. Soc. **67**, 97 (1955).
[2] See A. C. B. Lovell: Proc. Inst. Electr. Engrs. B **103**, 711 (1956).

Fig. 29. (Photo.) Large meridian telescope at Pott's Hill, Sydney. Its diameter is 35 feet, and at a wavelength of 21 cm it has a beamwidth to half power of 1.5 degrees.

Fig. 30. (Photo.) The Cambridge meridian interferometer which has four elements each 320 by 40 feet situated at the corners of a rectangle 1900 by 168 feet. The gain of each element is 950 at a wavelength of 3.7 meters.

6*

Fig. 31.

Fig. 32.

Fig. 33.

Fig. 31. (Photo.) A meridian array at Stanford. Length = 2020 feet, number of elements = 96, beamwidth to half power — 3 degrees at 23 MHz.

Fig. 32. (Photo.) Radio telescope on altazimuth mounting at Dwingeloo in the Netherlands. Diameter = 25 m, focal length = 12 m, beamwidth to half power at 21 cm = 0.56 degrees.

Fig. 33. (Photo.) The Manchester installation. Diameter = 250 feet, focal length = 62.5 feet.

Because radio telescopes are dish-shaped rather than pencil-shaped the various equatorial mountings for optical telescopes cannot be adopted unmodified. Most mountings abandon the feature of intersecting polar and declination axes, as a result of which the counterweights which bring the center of gravity of the reflector onto the declination axis have themselves in turn to be balanced by weights moving with the yoke to bring the center of gravity of the reflector plus yoke onto the polar axis. A further weight is then sometimes necessary to move the center of gravity along the polar axis to a point above the footing. A drawback associated with counterweights is the limit to motion imposed by their contact with the reflector, supporting structure, or ground; and modifications which increase the freedom of motion generally tend to increase the counterweights. A possibility occasionally adopted for small mountings is to make one of the counterweights variable with declination. Many ingenious variations can be worked out to suit special conditions, many no doubt still remaining to be invented. Fig. 35 shows some existing designs.

Fig. 34. (Photo.) One of a pair of 61-foot reflectors built by Stanford Research Institute, California.

48. Drive systems. Drive systems for tracking in sidereal time divide into those incorporating synchronous motors with suitable gearing, and others. Since the synchronous motor cannot change speed a gear change or second motor and differential gear must be provided for slewing. A remote position indicator, usually attached not to the reflector but to a gear wheel moving at higher speed, completes the system. A single non-synchronous motor can perform both tracking and slewing if the discrepancy between reflector time and clock time is used to control the motor speed. In one arrangement the clock turns the shaft of a synchro-receiver the error voltage from which excites one winding of a two-phase induction motor. In another the clock and reflector angles are subtracted mechanically and a switch turning the motor on and off is actuated by the difference shaft. An overriding control permits slewing at full speed.

Automatic returns from the western to the eastern horizon for long exposures on the one object, for example the Sun, require provision for keeping track of the object during the slewing. This is difficult when the synchronous element is the motor but is readily arranged when the synchronous element is the reference clock.

Special tracking rates such as required by the Sun, Moon, planets, etc., can be incorporated into non-synchronous drive systems by using appropriate reference clocks and into synchronous systems by adjusting the power frequency or by means of a further differential.

IV. Theory of aerial smoothing.

49. Definitions of brightness, brightness temperature, and flux density. The basic observation in radio astronomy is the determination of the strength of ratio waves coming from different directions over an area of sky. We may be concerned with a large part of the sky, as when making a galactic survey, or only a small part, as when studying the emission from discrete objects or sets of objects. We may also be concerned with the spectrum and the polarization and time dependence of strength, spectrum and polarization. In the case of those solar disturbances known as outbursts we have to deal with sources whose strength, spectrum, polarization, position and presumably size all change greatly in the course of minutes, but more usually the position is simpler and we begin with the measurement of strength as a function of position on the sky. This includes galactic surveys and the determination of position, size, and brightness distribution of discrete sources at a single frequency. All other types of measurement then reduce, at least in principle, to repetition on other frequencies, at other times, and with differently polarized aerials.

Three different quantities are customarily used in specifying the strength of celestial radio waves, viz. brightness and brightness temperature (in referring to extended sources) and flux density (in referring to discrete, but not necessarily point, sources). These quantities will now be defined.

Consider the energy ΔE which falls on a small area Δa at ground level in a time interval Δt and frequency band Δf, and which comes from directions within a cone of solid angle $\Delta \omega$ surrounding the point P on the celestial sphere through which passes the normal to the area. Let the epoch be t_1 and the mid-frequency f_1, and let us assume that there are no temporal, spectral, directional or spatial discontinuities[1] in the radiation at t_1, f_1, P or on the ground at Δa. Then we might expect that

$$\lim_{\Delta a, \Delta f, \Delta \omega, \Delta t \to 0} \frac{\Delta E}{\Delta a\, \Delta f\, \Delta \omega\, \Delta t}$$

would exist, thus defining a brightness b as a measure of the strength of radiation arriving at the receiving point from the direction of P and permitting us to write

$$dE = b\, da\, df\, d\omega\, dt.$$

In the meter-kilogram-second system of units, whose use in radio astronomy has been virtually universal, b is measured[2] in watts meters^{-2} Hz^{-1} steradians^{-1}. The precision with which one would know the strength of a signal occupying a band Δf and having a duration Δt is one part in $\sqrt{\Delta f\, \Delta t}$ for reasons already

---

[1] All these assumption can be expected to break down in practice in special cases.

[2] There is also an official movement under way to name this unit the jansky per steradian in honor of KARL JANSKY who first studied radio waves of extraterrestrial origin.

a

b

c

d

Fig. 35 a—d. (Photo.) A group of equatorially mounted radio telescopes. (a) The 60-foot reflector at Harvard (photo
by J. SHEAHAN, Boston Globe). (b) A model of the 140-foot reflector for Greenbank, West Virginia. (c) Christiansen
array at Fleurs, New South Wales, forming one arm of a cross. (d) One of thirty-two 10-foot reflectors at Stanford,
California.

explained in connection with the statistical limit to precision of noise measurement. Therefore it is necessary that $\Delta f\, \Delta t \gg 1$ and consequently it is not possible to proceed to the limit indicated above. Furthermore, the product $\Delta a\, \Delta \omega$ is subject to limitations, for suppose that the small collecting area Δa is to be realized by means of a horn or other aerial. Then $\Delta \omega$ will equal the "effective solid angle" $\lambda^2/\Delta a$ as defined earlier for an aerial of "effective area" Δa, and so the product $\Delta a\, \Delta \omega$ cannot be made less (or greater) than λ^2. This situation may not appear satisfactory as a basis for an observational science but it is accepted [1] and we can assert that the concept of brightness is useful in practice, the ratio $\Delta E/\Delta a\, \Delta \omega\, \Delta f\, \Delta t$ assuming a usably definite appearance while the factor $\Delta f\, \Delta t$ remains large enough to satisfy the inequality. The essential indefiniteness of b, as it is customarily conceived, reappears later.

Brightness is thought of as a scalar point function of direction (more generally a matrix of four quantities when polarization is included) specifying detail of the radiation field at a point. The use of brightness temperature T_b, as an alternative to b, derives part of its convenience from the proportionality at radio wavelengths between the brightness of the radiation from a black body and the temperature of the black body. Thus we may express Planck's radiation formula in the form

$$b = \frac{2 h f^3}{c^2}\, \frac{1}{\exp\left(\dfrac{h f}{k T}\right) - 1},$$

where $b\, df$ is the power in the frequency range f to $f + df$ which is received perpendicularly per steradian on one side of unit area situated in an isotropic radiation field at temperature T [2]. When $h f \ll k T$, which is true for radio wavelengths

---

[1] In radiation theory, for example in the thermodynamics of stellar atmospheres, the concept of brightness is introduced along the lines of Milne's well known treatment (Handbuch der Astrophysik, vol. 3, p. 65, 1930). Milne did not mention the restriction on Δt and it continues to be overlooked in recent treatises (of which for example see A. Unsöld, Physik der Sternatmosphären, Berlin 1955); however it is known in connection with the description of signals in the time-frequency domain [see D. Gabor, J. Inst. Electr. Engrs. III **93**, 429 (1946), who also gives earlier references], and is an essential mathematical property according to which Δf and Δt, defined as the standard deviations of the spectral and temporal energy distributions about their mean abscissae, cannot, for a wide variety of signal functions, have a product less than $1/4\pi$. This mathematical fact assumes various embodiments in different branches of physics. Thus in quantum physics it appears in Heisenberg's uncertainty relation [W. Pauli, this Encyclopedia vol. V/1, p. 20 (1957)], and in classical diffraction theory implies that $\Delta a'\, \Delta \omega' \geqq \lambda^2/(4\pi)$, where $\Delta a'$ and $\Delta \omega'$ are suitably redefined in terms of the standard deviations of the aperture intensity distribution and its power radiation pattern. This condition, together with the condition $\Delta f\, \Delta t \gg 1$, plays an important role in assessment of the efficiency of, and in setting a limit to, the rate of acquisition of information by an instrument. [In the strict proof that $\Delta f\, \Delta t \geqq 1/(4\pi)$, one defines Δf to be the standard deviation of the squared modulus of the Fourier transform of the signal waveform. For a quasi-monochromatic wave packet, $\Delta f \sim f$, and the inequality, through true, fails to be practical by a huge factor. If one defines Δf in the more natural way as the standard deviation of the positive-frequency part of the squared modulus of the transform of the signal, then $\Delta f\, \Delta t$ no longer possesses a strict lower limit of $1/(4\pi)$, as has been shown by E. Wolf, Proc. Phys. Soc. Lond. **71**, 257 (1958) and by I. Kay and R. A. Silverman, Information and Control **1**, 64 (1957).]

[2] In m.k.s. units we have

Planck's constant, $h = 6.623 \times 10^{-34}$ joule-seconds,

Boltzmann's constant, $k = 1.3803 \times 10^{-23}$ joules degrees^{-1},

velocity of light, $c = 2.99792 \times 10^8$ meters seconds^{-1},

and

$$b = \frac{3.97 \times 10^{-25}\, \lambda^{-3}}{\exp\left(0.0143/\lambda T\right) - 1}.$$

except at the very lowest temperatures, we have the Rayleigh-Jeans approximation

$$b = \frac{2kT}{\lambda^2}.$$

This equation defines a temperature for every b, irrespective of whether the radiation field is of thermal origin or not, and this temperature is proportional to power. As we have seen in connection with the calibration procedure for receivers, absolute measurements are commonly made by comparison with the thermal radiation from a resistor at known temperature, the so-called available power $kT\,\Delta f$ of a resistor being itself an expression of the Rayleigh-Jeans approximation. Brightness temperature T_b is thus not only a convenient alternative to brightness b, for it is more often the actual datum.

The last of the three customary quantities, the flux density S of a source, may be defined as

$$S = \iint b\,d\omega,$$

where the integral is taken over the whole solid angle subtended by the source and b is that part of the total brightness at the receiving point which is deemed due to the "source". Flux density is thus a scalar point function of position, as is b, but is not a function of direction. It is however a function of the possibly arbitrary outline deemed to define the source and therefore involves a subjective judgment by an observer. The way in which the subjective element enters is seen in a later section on the determination of flux density.

There is a fourth quantity, apparent disc temperature T_D, defined by

$$\frac{2kT_D}{\lambda^2} = \frac{S}{\Omega_D},$$

which is applicable to objects such as the Sun and planets which have definite optical discs or any other characteristic solid angle Ω_D. Since the optical disc may be unrelated to the size of the radio source (for example consider the corona and plages), the apparent disc temperature has deficiencies. But it is apposite, and has proved useful, in the absence of knowledge about the radio source.

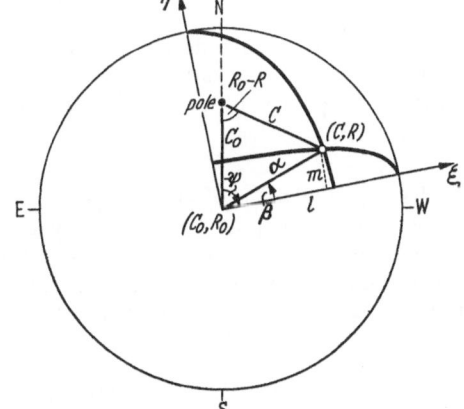

Fig. 36. Orthogonal projection of the sky on the aperture plane.

50. The basic convolution relation. Let $T(C, R)$ be the distribution of true brightness temperature for the frequency and polarization accepted by an aerial, C and R being codeclination and right ascension respectively.

To specify the orientation of the aerial it is necessary to give not only (C_0, R_0), the direction of the beam axis, but also a position angle ψ which determines the rotation of the beam about this axis. The position angle ψ gives the direction of a transverse axis, fixed in the aerial, measured eastwards from north (Fig. 36).

$D(\alpha, \beta)$ is the directivity of the aerial in the direction (α, β), α and β being spherical polar coordinates relative to axes fixed in the aerial. The polar angle α is measured from the ζ-axis; the longitude β in the same sense as ψ from the great circle containing the ζ- and ξ-axes. From the definition of directivity, $D(\alpha, \beta)$ is normalized so that

$$\int_0^{2\pi} \int_0^{\pi} D(\alpha, \beta) \sin\alpha \, d\alpha \, d\beta = 4\pi.$$

We wish to calculate the available power, as defined in Sect. 26, when the aerial is pointed towards (C_0, R_0) with position angle ψ. This we can do with the aid of the principle of detailed balancing, quoting from Pawsey and Bracewell[1]. "Consider the situation of Fig. 37 in which it is required to find the available power at the terminals of aerial A due to thermal radiation from the body M which is at one temperature T. Let us tentatively connect to the terminals an impedance Z, for which the power transfer is maximum, and make the temperature of Z and of all the surroundings equal to T so that thermodynamic equilibrium may be realized. Then by the principle of detailed balancing the energy in the frequency range Δf transferred from M to Z equals that transferred from Z to M. Suppose a fraction α of the total radiation from the aerial is absorbed in M. The thermal power delivered by Z to the aerial is $kT\,\Delta f$, so that $\alpha kT\,\Delta f$ is absorbed in M. This quantity is also transferred from M to A. If Z is changed the actual power transfer may be altered but the available power from M will be as before. Also the temperature of the surroundings other than M may depart from T without affecting the contribution from M itself (if the changes do not alter its physical state or attenuate or deviate the radiation). The total available power P from all sources will be the sum of the contributions from all surrounding bodies. This is conveniently expressed in terms of the effective aerial temperature T_a, where

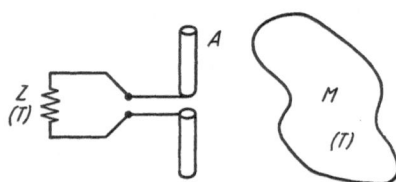

Fig. 37. An impedance Z in thermodynamic equilibrium with its surroundings at temperature T.

$$P = k\,\Delta f\,T_a = k\,\Delta f \sum \alpha_n T_n,$$

and the summation extends over all the surrounding bodies, α_n and T_n being the fraction of power radiated from the aerial absorbed in the body and the temperature, respectively, of the n-th body. This reduces to

$$T_a = \sum \alpha_n T_n,$$

so that T_a is a mean of the temperature of the surroundings weighted according to the fraction of power radiated by the aerial absorbed in each."

Let the direction (C, R) be the same as (α, β) when the aerial is placed in position angle ψ with its ζ-axis pointing in the direction (C_0, R_0). Then the fraction α of radiated power which would be absorbed by a black body subtending a solid angle $\sin C \, dC \, dR$ in the direction (C, R) is

$$D(\alpha, \beta) \frac{\sin C \, dC \, dR}{4\pi}.$$

Hence the effective aerial temperature T_a of a loss-free aerial surrounded by black bodies of temperature $T(C, R)$ in the direction (C, R) *or* surrounded by a brightness temperature distribution $T(C, R)$, is given by

$$T_a = \frac{1}{4\pi} \iint T(C, R) \, D(\alpha, \beta) \sin C \, dC \, dR.$$

[1] J.L. Pawsey and R.N. Bracewell: Radio Astronomy. Oxford 1955.

This integral may be taken over all directions in space or, with advantage, over the sky only, $D(\alpha, \beta)$ then being interpreted as the directivity of the aerial plus Earth. Since we have assumed that the total available power from all directions will be the sum of the separate contributions the conclusion will be invalid if there is coherence, such as would be produced by ground reflections, between rays arriving from different directions; and a good way to validate it is to deem the Earth to be part of the aerial.

Before the integral can be evaluated it is necessary to know C and R in terms of C_0, R_0, α, β, and ψ, for example by use of the relations

$$\cos C = \cos C_0 \cos \alpha + \sin C_0 \sin \alpha \cos (\beta + \psi)$$

$$\cot (R_0 - R) = \cos C_0 \cot (\beta + \psi) - \sin C_0 \cot \alpha \operatorname{cosec} (\beta + \psi).$$

We have now derived the basic formula of aerial smoothing in radio astronomy. The principal assumption, incoherence of the sources of radiation, is expected to prove widely valid for celestial sources, though it is also to be expected that the sky may contain examples of coherence caused by refraction or diffraction.

Before we can proceed further it is necessary to consider whether the function $D(\alpha, \beta)$ is unchanged when the ζ-axis points to different parts of the sky. In a great many practical cases it is accurate to assume that this is so, but there are important cases where it is not, for example movable aerials receiving significant ground reflections, and interferometers with aerials rotatable about axes not parallel to a line containing the elements. When the Earth is deemed to be part of the aerial as before, all these cases appear as non-rigid aerials whose parts may possess relative motion, and in the following development these cases are excluded; but the exclusion does not extend to the above aerials when they are caused to scan purely by the Earth's rotation.

It is desirable to specialize the basic equation to a simpler form and to distinguish two major branches of the theory according as ψ is retained or abandoned. By assuming that $\psi = \text{const}$ we restrict attention to radiation patterns subject only to "parallel" displacements; by allowing ψ to vary we generate the subject of strip integration which is taken up again in a later section.

Now placing the aerial on an equatorial mounting (an altazimuth mounting will allow ψ to vary) and causing the ξ- and η-axes of the aerial to fall respectively along declination and hour circles ($\psi = -\frac{1}{2}\pi$) we note that $-R \operatorname{cosec} C$ and C are approximate rectangular coordinates of points lying in a zone of declination containing the beam axis which is not too wide and not near the poles. Calling these coordinates x and y, and restricting attention to highly directional aerials within a certain zone of declination, we finally reduce the aerial smoothing equation to

$$T_a(x, y) = \iint A(x', y') \, T(x + x', y + y') \, dx' \, dy',$$

where x' and y', the coordinates of the element $dx' \, dy'$ relative to the ζ-axis of the aerial are essentially the same as l and m, $T(x, y)$ and $T_a(x, y)$ are the true and observed brightness temperatures respectively, and $A(x', y')$ is proportional to the directivity of the aerial but so normalized that

$$\iint A(x', y') \, dx' \, dy' = 1.$$

In these double integrals and those that follow, infinite limits are understood, but by the assumption of highly directional aerials the integrand differs from zero over only a small range of variables.

It will be noticed that in the notation introduced earlier (Sect. 43)

$$T_a = A \star T = \iint A(x', y') \, T(x + x', y + y') \, dx' \, dy'$$
$$= \iint A(x' - x, y' - y) \, T(x', y') \, dx' \, dy'.$$

We may force this result into the form of a convolution integral (denoted by $*$) by working in terms of A, the response to a point source $^2\delta(x, y)$, instead of the conventional radiation pattern A. Substituting above,

$$A(x, y) = A \star {}^2\delta = A(-x, -y).$$

Then

$$T_a = A * T = T * A = \iint A(x', y') \, T(x - x', y - y') \, dx' \, dy'$$
$$= \iint A(x - x', y - y') \, T(x', y') \, dx' \, dy'.$$

Because the operation of convolution is commutative ($f * g = g * f$), associative ($f * [g * h] = [f * g] * h$), and distributive ($[f + g] * h = f * h + g * h$) (see Doetsch[1]), it may be treated algebraically like ordinary multiplication and thus leads to simple mathematics, as below. On the other hand the operation of smoothing T with A (written $A \star T$) is perhaps more direct for some purposes than forming the convolution $A * T$. In either case it is A which is plotted or tabulated when the calculation is performed. And, of course, A is the customary quantity in aerial physics, not A. However, the smoothing process, being non-commutative and non-associative, proves to be not as convenient as convolution in the type of analysis presented below. But when the radiation pattern is symmetrical, the distinction between A and A disappears.

51. Formal solution by Fourier transforms. The Fourier transforms of functions related by convolution, as when

$$T_a = A * T,$$

themselves have a simple product relationship, viz.

$$\overline{T}_a = \overline{A} \, \overline{T},$$

where bars represent Fourier transforms, i.e.

$$\overline{T}(u, v) = \iint e^{-i 2\pi (u x + v y)} \, T(x, y) \, dx \, dy,$$

and conversely

$$T(x, y) = \iint e^{i \pi (u x + v y)} \overline{T}(u, v) \, du \, dv.$$

This is the two dimensional convolution theorem of which the array theorem introduced earlier is an expression.

In one dimension we write

$$\overline{T}(s) = \int e^{-i 2\pi s x} \, T(x) \, dx$$

and we shall often draw on this simpler form for illustration in what follows. The quantity s is the number of cycles per unit of x, or the spatial frequency, of a Fourier component of $T(x)$. In two dimensions a single Fourier component $e^{-i 2\pi (u x + v y)}$ may be regarded as a train of parallel crests and troughs in the (x, y) plane, proceeding in a direction inclined at an angle $\arctan(v/u)$ to the x-axis, with a spatial frequency $\sqrt{u^2 + v^2}$ cycles per unit distance in the xy-plane.

---

[1] G. Doetsch: Theorie und Anwendung der Laplace-Transformation. Berlin 1937.

The spatial frequency of a cross-section parallel to the x-axis is u, and parallel to the v-axis is y. We regard the true distribution $T(x, y)$ as composed of waves proceeding in all directions with all spatial frequencies, each wave of appropriate strength $T(u, v)$. Then the product formula tells us that $\overline{T}_a$ is derived from $\overline{T}$ by multiplication with a factor $\overline{A}$. Any values of $\sqrt{u^2 + v^2}$ for which $\overline{A}$ is zero are particularly important, because then the modification is complete rejection. Knowing $\overline{T}_a$ and $\overline{A}$, we can partially infer $\overline{T}$; thus

$$\overline{T} = \frac{\overline{T}_a}{\overline{A}}$$

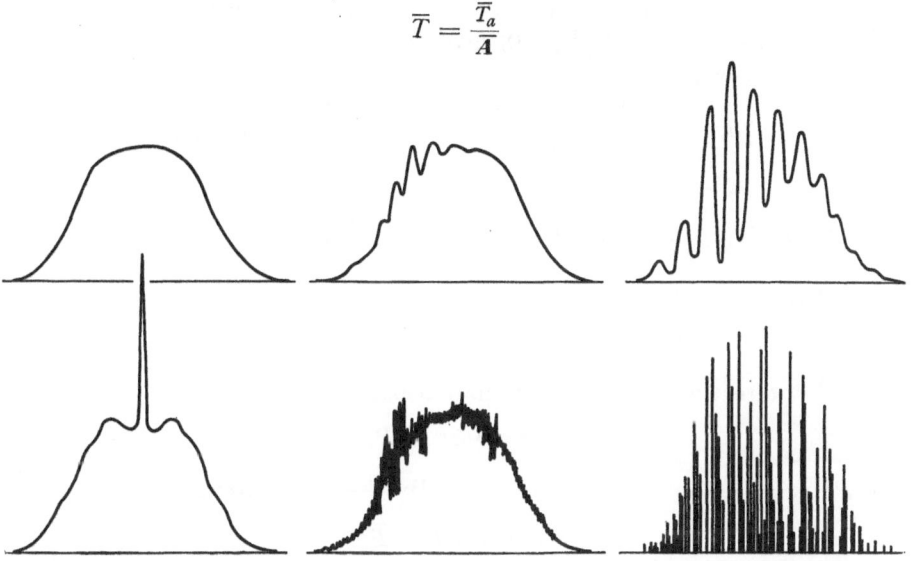

Fig. 38. A variety of distributions which when scanned with the same aerial all give the same result.

for values of $\sqrt{u^2 + v^2}$ such that $\overline{A} \neq 0$. For other values of spatial frequency we can say nothing. Let $\overline{A}$ be zero where $u = u_k$, $v = v_k$. Then the product equation will be satisfied not only by $\overline{T}$ but also by

$$\overline{T} + \sum_k a_k \, {}^2\delta(u - u_k, v - v_k),$$

where the coefficients a_k are arbitrary and ${}^2\delta(u - u_k, v - v_k)$ is a unit two-dimensional impulse at $u = u_k$, $v = v_k$. It follows that T is not the only solution of the convolution equations; it is also satisfied by

$$T + \sum_k a_k \, e^{i 2\pi (u_k x + v_k y)}.$$

(If $\overline{A}$ is zero, not at discrete points, but over a continuous range, the summation is replaced by an integration.) The additive functions

$$\sum_k a_k \, e^{i 2\pi (u_k x + v_k y)},$$

which we term invisible distributions for the aerial, are obviously solutions of the integral equation

$$A \star T = 0.$$

They are of such a nature that it is impossible to detect them with the aerial in question, whatever their magnitude. Some one-dimensional examples of distributions containing invisible components are given in Fig. 38 taken from a paper by Bracewell and Roberts[1] which first discussed this aspect of the non-uniqueness of the solution of the aerial smoothing equation.

It is possible to use the formal solution in practice but only in simple cases and then only with caution and often with disappointing results for numerical reasons. However, as a theoretical basis for further discussion it is invaluable. It is clear that the zeros of $\bar{A}$ play a vital role and we now investigate from this point of view the Fourier transform of aerial patterns.

52. The spectral sensitivity theorem. Let an aerial specified by an electric field distribution $E_\eta = E(\xi/\lambda, \eta/\lambda)$, $E_\xi = 0$, give rise to an angular spectrum $P(l,m)$. Then we have shown that

$$P(l, m) = \iint E\left(\frac{\xi}{\lambda}, \frac{\eta}{\lambda}\right) e^{-i 2\pi \left(l\frac{\xi}{\lambda} + m\frac{\eta}{\lambda}\right)} d\left(\frac{\xi}{\lambda}\right) d\left(\frac{\eta}{\lambda}\right)$$

and

$$E\left(\frac{\xi}{\lambda}, \frac{\eta}{\lambda}\right) = \iint P(l, m) e^{i 2\pi \left(l\frac{\xi}{\lambda} + m\frac{\eta}{\lambda}\right)} dl\, dm.$$

We may write this latter equation

$$E = \bar{P}.$$

For a highly directional aerial the radiation pattern A is given by

$$A = \text{const } P^* P,$$

where P^* is the complex conjugate of P. By the convolution theorem

$$\bar{A} = \text{const } \overline{P^*} * \overline{P}$$
$$= \text{const } E^*(-) * E$$
$$= \text{const } E^* \star E,$$

hence,

$$\bar{A} = \text{const } E \star E^*.$$

In this proof we have used the fact that the transform of the conjugate of P, denoted by $E^*(-)$, is obtained from E^*, the conjugate of its transform, by reversing the signs of ξ and η.

The quantity $\bar{A}$ which we are investigating, and which we may call the spectral sensitivity function, is thus proportional to the complex autocorrelation function of E. The constant of proportionality follows from the normalization of $\bar{A}$; since $\iint A\, dl\, dm = 1$ it follows that $\bar{A}|_0 = 1$. Therefore, dividing $E \star E^*$ by its value for $u = v = 0$, we find finally that the spectral sensitivity function of an aerial is equal to the complex autocorrelation function of its aperture distribution, normalized in the conventional way, i.e. with a central value of unity:

$$\bar{A}(u, v) = \frac{\iint E\left(\frac{\xi}{\lambda} - u, \frac{\eta}{\lambda} - v\right) E^*\left(\frac{\xi}{\lambda}, \frac{\eta}{\lambda}\right) d\left(\frac{\xi}{\lambda}\right) d\left(\frac{\eta}{\lambda}\right)}{\iint E\left(\frac{\xi}{\lambda}, \frac{\eta}{\lambda}\right) E^*\left(\frac{\xi}{\lambda}, \frac{\eta}{\lambda}\right) d\left(\frac{\xi}{\lambda}\right) d\left(\frac{\eta}{\lambda}\right)}.$$

We may now restate the basic relation of aerial smoothing in the form

$$\bar{T}_a = \text{const } (E \star E^*)\, \bar{T}.$$

---

[1] R.N. Bracewell and J.A. Roberts: Austral. J. Phys. **7**, 615 (1954).

This is an expression of the spectral sensitivity theorem, i.e. it states the relation between the transforms of T_a and T in terms of the aerial aperture distribution E.

53. The aerial cut-off theorem. An aerial theorem of far-reaching consequences may now be proved. Since aerials are finite in extent the function E falls to zero for values of ξ (or ξ and η) greater than the finite values corresponding to the extremities of the aerial. When the function $E \star E^*$ is evaluated, the result, namely $\overline{A}$, must have this same property of falling to zero beyond a finite region; for if we examine the numerator in the expression for $\overline{A}$, we see that one or other of the factors is zero for all ξ and η when u and v exceed certain values depending on the exact extent of E over the $\xi\eta$-plane. A graphical procedure is described in Sect. 59.

For illustration consider an aerial consisting of a finite one-dimensional aperture of width w, across which is maintained a field constant in amplitude and phase. For this aerial the $A(x)$ for small x is approximately

$$A(x) = \frac{\lambda}{w} \left[\frac{\sin\left(\frac{\pi x w}{\lambda}\right)}{\pi x} \right]^2 ,$$

the numerical factor being chosen so that $\int A(x)\,dx = 1$. Fig. 39 shows $A(x)$ and $\overline{A}(s)$, which can conveniently be plotted against $|s|$ in this case since $\overline{A}(s)$ is real and even. The beam width of the main

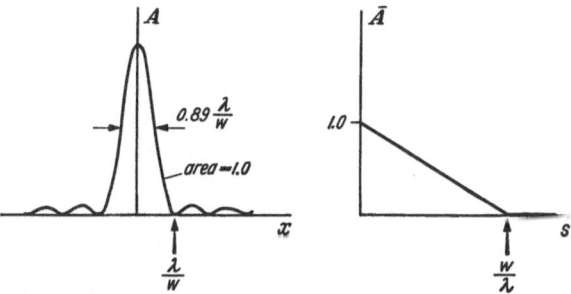

Fig. 39. The response to a point source, $A(x)$, and its Fourier transform $\overline{A}(s)$, for a uniformly excited aperture.

lobe between zeros is $2\lambda/w$ and the width to half power is $0.89\,\lambda/w$. The particular feature to note is that $\overline{A}(s)$ is zero for all values of s greater than the limiting spatial frequency $s_c = w/\lambda$.

The consequences of $\overline{A}$ falling to zero in this as in all other cases[1], is that $\overline{T}_a$ necessarily does the same, being derived from $\overline{T}$ by multiplication with $\overline{A}$. A powerful discrete-interval theorem then applies.

54. The discrete-interval theorem. A function $T_a(x, y)$ such that $\overline{T}_a(u, v)$ is zero for $|u| \geq u_c$ or $|v| \geq v_c$, is completely determined by its values at the points $(m/2u_c' + a,\, n/2v_c' + b)$, where m and n assume all integral values, a and b are arbitrary constants, and the spacing between points may be as wide as is compatible with $u_c' \geq u_c$ and $v_c' \geq v_c$.

The condition on $\overline{T}_a(u, v)$ may be expressed by saying that it is zero on and outside a rectangle which is centered on the origin of the uv-plane and set with its sides parallel to the axes; and we may note that this covers the case of $\overline{T}_a(u, v)$

---

[1] Our proof applies strictly only to plane apertures, but we know from experience that the directivity of a highly directional array cannot be improved much by rearranging its elements within the same overall dimensions. In any particular case the existence of the cut-off could be verified, and the value of s_c determined, by taking the Fourier transform of the radiation pattern. Low-gain apertures which are comparable with, or less than, a wavelength in extent, can be harmonized with the present treatment by considering only that part of the aperture distribution which does not generate evanescent fields. In fact strictly speaking the function $E \star E^*$ should be evaluated in all cases only after removal from E of those components which do not radiate. By not doing so one ignores an effective extension of the aperture distribution to a distance of about $\lambda/2$. To this extent, an aperture distribution never really cuts off sharply.

zero on and outside a circle or other region provided the rectangle is chosen sufficiently large.

It is sufficient to give a proof for the case where a and b are zero, i.e., where the origin of x and y is one of the sampling points. For, if the transform of $T_a(x, y)$ is zero on and outside a given rectangle, so also is that of $T_a(x+a, y+b)$ by virtue of the two-dimensional shift theorem, according to which the Fourier transform of $T_a(x+a, y+b)$ is $\overline{T}_a(u, v) \exp(i\, 2\pi(au+bv))]$, which must be zero. Therefore, if the theorem is true for $T_a(x, y)$, it is also true for $T_a(x+a, y+b)$; but values of $T_a(x+a, y+b)$ at points of an array which includes the origin are values of $T_a(x, y)$ taken over an offset array.

To prove the theorem we use the bed-of-nails function $^2III(x, y)$ consisting of a two-dimensional array of unit impulses separated by unit distance. Thus

$$^2III(x, y) = \sum_{m=-\infty}^{\infty} \sum_{n=-\infty}^{\infty} {}^2\delta(x-m, y-n).$$

The bed-of-nails function is known to be its own two-dimensional Fourier transform (Bracewell[1]).

Proof of theorem. Let $F(u, v) \equiv (4u_c' v_c')^{-1}\, ^2III(u/2u_c') * \overline{T}_a$, a function which may be pictured as an array of islands in the uv-plane, each the same as $\overline{T}_a$, spaced at intervals $2u_c'$ in the u direction and $2v_c'$ in the v direction. The islands will not overlap (but may touch) if $u_c' \geqq u_c$ and $v_c' \geqq v_c$. Under this condition, in the central region where $|u| < u_c$ and $|v| < v_c$, we have

$$\overline{F}(u, v) = \overline{T}_a.$$

Hence $\overline{T}_a$ may be recovered from $\overline{F}(u, v)$, and consequently T_a may be recovered from $F(x, y)$, the two-dimensional Fourier transform of $\overline{F}(u, v)$. But, by the two-dimensional convolution theorem,

$$F(x, y) = 4u_c' v_c'\, ^2III(2u_c' x, 2v_c' y),$$

which contains values of T_a only at discrete intervals $(2u_c')^{-1}$ and $(2v_c')^{-1}$ of x and y. Hence T_a is completely determined by its values at discrete intervals of x and y which are equal to or less than $(2u_c)^{-1}$ and $(2v_c)^{-1}$. Since these intervals are peculiar to each aerial they will be referred to as the peculiar intervals.

This theorem is of great importance. For observational work it means that observations need not be more closely spaced than a limit determined by the aerial pattern. For computational work the property is equally important as it permits observed data to be represented exactly by a set of discrete values. An inverse form of the theorem, relating to celestial sources of finite angular extent, means that observations of coherence need not be made at spacings less than a limit set by the source size.

V. Interferometers.

a) Two-element interferometers.

As used in radio astronomy the term interferometer signifies an aerial with two or more well separated parts. When the aerial is used to transmit, the fields from the different parts combine in the distance to form spatial interference fringes and when it is used to receive from a moving point source the received power undergoes temporal variations which themselves are often called fringes. No confusion is caused by this transference of nomenclature, as the spatial fringes as such are not observed.

[1] R.N. Bracewell: Austral. J. Phys. **9**, 297 (1956).

Since interferometers are aerials, they are covered by the aerial theory which we have developed; they are distinguished from aerials in general by the possession of more or less periodic radiation patterns, a feature associated with two or more well separated parts.

55. Monochromatic theory. Consider first two identical aerials with directivity $D(\vartheta, \varphi)$ connected symmetrically together as shown in Fig. 40. Then the directivity of the combination can be deduced as follows. At a distant point in the direction (ϑ, φ), where ϑ and φ are the same angles as those in Fig. 18, two field components will combine with a phase differ-

ence $\delta = \dfrac{2\pi}{\lambda}\, a \sin \vartheta$ and amplitudes which are equal but $1/\sqrt{2}$ times less than would have been produced by a single aerial radiating as much total power as the interferometer. The flow of power into the direction (ϑ, φ) is therefore

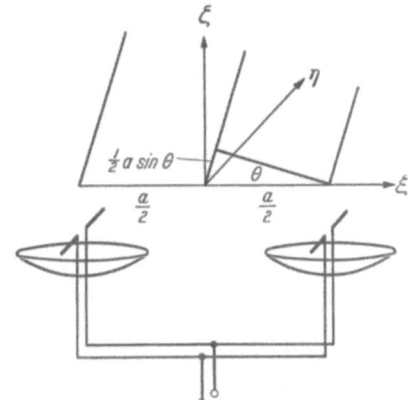

$$\left(\frac{1}{\sqrt{2}}\, 2 \cos \frac{1}{2}\, \delta\right)^2 = 1 + \cos \delta$$

relative to a single aerial and so the directivity of the interferometer is

$$D_i(\vartheta, \varphi) = D(\vartheta, \varphi)\left(1 + \cos \frac{2\pi a \sin \vartheta}{\lambda}\right).$$

Fig. 40. A two-element interferometer.

It will be noticed that the modifying factor is independent of φ and that the loci of constant ϑ are small circles with centers on the ξ-axis. For example a two element interferometer lying on an east-west line has a radiation pattern

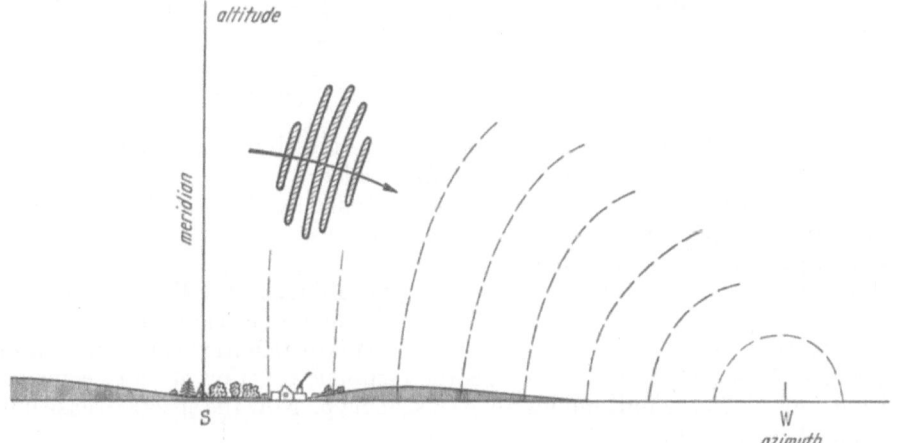

Fig. 41. View of the south-western horizon. The shading represents the reception pattern of a two-element east-west interferometer.

as illustrated in Fig. 41 which shows a view of the sky looking south. In this type of diagram, which the author commends for explanatory purposes, we shade those areas of sky where the directivity exceeds half the maximum. The envelope of the fringe system is the half-directivity diagram of a single aerial, and may be steered about in the sky by steering the aerials; but this does not move the fringes from their location on the small circles having the west point of the horizon as pole.

If a point source were to move through the reception pattern along the arrow shown in Fig. 41 the time variation of received power would be as shown in Fig. 42. If a single aerial were used the response would be the broken line.

Let the directivity $D(\vartheta, \varphi)$ and angular spectrum P be those which would result from an extended aperture distribution $E\left(\frac{\xi}{\lambda}, \frac{\eta}{\lambda}\right)$. Then a pair of such apertures may be represented by $II * E$, where II is a pair of impulse functions,

$$II = {}^2\delta\left(\frac{\xi}{\lambda} - \frac{a}{2\lambda}, \frac{\eta}{\lambda}\right) + {}^2\delta\left(\frac{\xi}{\lambda} + \frac{a}{2\lambda}, \frac{\eta}{\lambda}\right).$$

The angular spectrum P_i of the interferometer is given by

$$P_i = \overline{II * E}$$
$$= \overline{II}\ \overline{E}$$
$$= 2\left(\cos\frac{\pi a l}{\lambda}\right) P.$$

Then

$$P_i P_i^* = 4\left(\cos^2\frac{\pi a l}{\lambda}\right) P P^*$$

and

$$D_i(\vartheta, \varphi) = D(\vartheta, \varphi)\left(1 + \cos\frac{2\pi a \sin\vartheta}{\lambda}\right).$$

This alternative derivation of the directivity of an interferometer illustrates an algebraic approach which is often useful for thinking out new systems. In

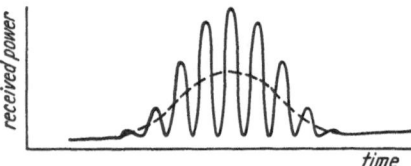

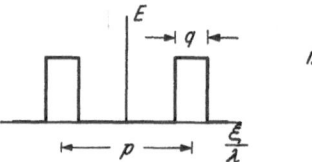

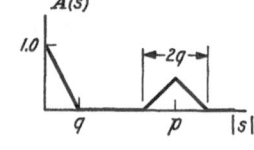

Fig. 42. The power received from a point source passing through the reception pattern of Fig. 41.

Fig. 43. Aperture distribution of an interferometer and its spectral sensitivity function.

effect we have said that a new system has been derived by convolution of a simple aperture distribution with a double impulse, which we know to be the Fourier transform of a cosine wave in the l-direction. Hence by the convolution theorem the old angular spectrum must be multiplied by a cosine variation and the old radiation pattern by cosine squared.

We now derive, under the conditions of applicability of the Fourier transform formula, the spectral sensitivity function for the one-dimensional interferometer whose aperture distribution is shown in Fig. 43. This distribution can be recognized as the convolution $II_q * II_p$ where II_q is the rectangle function of unit height and width q, and II_p is the unit impulse-pair of spacing p. Now the spectral sensitivity function $\overline{A}$ is given by

$$\overline{A} = \text{const } E \star E^*$$
$$= \text{const } E * E$$
$$= \text{const } (II_q * II_p) * (II_q * II_p)$$
$$= \text{const } (II_q * II_q) * (II_p * II_p)$$
$$= \Lambda_q * {}_1 II_1$$
$$= \tfrac{1}{2}\Lambda_q(s + p) + \Lambda_q(s) + \tfrac{1}{2}\Lambda_q(s - p),$$

where $\Lambda_q = \Pi_q * \Pi_q$ is the triangle function of unit height and width $2q$, $_I I_I =$
$\frac{1}{2}\delta(s+p) + \delta(s) + \frac{1}{2}\delta(s-p)$, and the constant has been adjusted to make $\bar{A}(0) = 1$.
The important feature to notice is that the interferometer responds to a band
of Fourier components of spatial frequency centered on p and of total width
q, where p is the number of wavelengths between the centers of the elements
and q is the breadth of each element in wavelengths. In addition, the interferometer responds to uniform brightness ($s=0$) and to low-frequency spatial components up to q cycles per radian. The curve of Fig. 42 contains just such bands
of Fourier components, one at low-frequencies and one centered on a high frequency; in fact, since the spectral sensitivity function (Fig. 43) and the response
to a point source Fig. 42 are a Fourier transform pair they afford precisely equivalent descriptions of the behavior of an interferometer.

Having derived the properties of a two-element interferometer and considered
a simple example, we calculate what is observed when an interferometer is used
on a source distribution $T(x-Vt, y)$ which is moving with an angular velocity V
in the x-direction past an interferometer whose radiation pattern is $A(x, y) \times$
$(1+\cos 2\pi S x)$. The aerial temperature T_a will be given by

$$T_a = \iint T(x - Vt, y) A(x, y) (1 + \cos 2\pi S x) \, dx \, dy$$
$$= T_{a_1}(-Vt) + \cos 2\pi S Vt \iint \cos 2\pi S x \, T(x, y) A(x + Vt, y) \, dx \, dy -$$
$$- \sin 2\pi S Vt \iint \sin 2\pi S x \, T(x, y) A(x + Vt, y) \, dx \, dy.$$

The term $T_{a_1}(-Vt)$ is what would be received on a single aerial. The remainder
terms take on a simple form if the distribution T is compact relative to A and is
mainly concentrated around (x_1, y_1). Then

$$T_a \approx A(x_1 + Vt, y_1) \left[\iint T(x, y) \, dx \, dy + \right.$$
$$+ \cos 2\pi S Vt \iint \cos 2\pi S x \, T(x, y) \, dx \, dy - \sin 2\pi S Vt \iint \sin 2\pi S x \, T(x, y) \, dx \, dy \right]$$
$$= A(x_1 + Vt, y_1) \iint T(x, y) \, dx \, dy \left[1 + N \cos(2\pi S Vt - \beta) \right],$$

where

$$N e^{i\beta} = \frac{\iint e^{-i 2\pi S x} T(x, y) \, dx \, dy}{\iint T(x, y) \, dx \, dy}.$$

The remainder thus approximates to an oscillation whose phase and amplitude
are given by that Fourier component of $T(x, y)$ having S cycles per unit of x
in the x-direction, the whole modulated by the radiation pattern centered on
the time of closest approach of (x_1, y_1) to the beam axis. The quantity N which
measures the amplitude of oscillation relative to the mean is what in optics would
be called the visibility of the fringes. By returning to the simple one dimensional
illustration we can illustrate this Fourier-transforming action of an interferometer.

As the width q of the individual aerials approaches zero the aperture distribution can be represented by II and the sensitivity function by $_I I_I$. Hence the
elementary aerial-pair spaced p wavelengths responds to source components
of two spatial frequencies: (i) p cycles per radian, (ii) zero cycles per radian.

Since any aperture distribution can be regarded as composed of numbers of
infinitesimally wide elements, it follows that an aerial will respond to all those
Fourier components of spatial frequency p' such that the aperture contains a
pair of elements p' wavelengths apart. Thus in the example of Fig. 43 the aperture distribution does not excite any two points whose spacing is greater than q
and less than $p-q$, nor any whose spacing exceeds $p+q$. This is reflected by
absence of response in the graph of spectral sensitivity function.

This concept of spaced pairs of elements is a very useful one and it also gives the strength of response. One simply counts the numbers of ways in which a pair of spacing p' can be found, allowing due weight for strength of excitation. Thus again referring to Fig. 43, the spacing $p - q + \varepsilon \, (\varepsilon < q)$ can be found in a number of ways proportional to ε. All the linear segments of the graph of $\overline{A}$ can be explained in this way. Furthermore the height of the maximum at $s = 0$ must be twice that of the maximum at $s = p$, for zero spacing can be found in twice as many ways as the spacing p. To prove these statements, which have been worded with a certain lack of rigor, one simply refers to the integral expression for $\overline{A}$ namely $E \star E^*(s)$, which prescribes just how the elements of spacing s are to be "counted" with "due weight". This view of the matter translates directly into two dimensions (Sect. 65) and throws immediate light on the effect of bandwidth (Sect. 57).

56. Coherence. We have measured the power arriving from the direction (x, y) in terms of a temperature $T(x, y)$, but we know that the instantaneous power is subject to random fluctuations. If attention is concentrated on a narrow band of frequencies, the distant field in the direction (x, y) may be represented by a complex phasor $f(x, y)$ whose modulus will have a Rayleigh distribution and whose phase drifts through all values equally. Let

$$\langle f f^* \rangle = T,$$

where

$$\langle \cdots \rangle \equiv \lim_{L \to \infty} \frac{1}{2L} \int\limits_{-L}^{L} \cdots dt.$$

Since an interferometer composed of two small aerials at a certain spacing can measure one Fourier component of the distribution $T(x, y)$, it is in fact measuring one spatial component of $\langle f f^* \rangle$. The field distribution $f(x, y)$ produces in the $\xi\eta$-plane a Fraunhofer diffraction field $F\left(\frac{\xi}{\lambda}, \frac{\eta}{\lambda}\right)$ such that

$$F\left(\frac{\xi}{\lambda}, \frac{\eta}{\lambda}\right) \propto \iint f(x, y) \, e^{-i 2\pi \left(x \frac{\xi}{\lambda} + y \frac{\eta}{\lambda}\right)} dx \, dy.$$

Hence the Fourier transform of $f f^*$ will be proportional to the spatial complex autocorrelation of the phasor F:

$$\iint f(x, y) \, f^*(x, y) \, e^{-i 2\pi \left(x \frac{\xi}{\lambda} + y \frac{\eta}{\lambda}\right)} dx \, dy$$
$$\propto \iint F(\alpha, \beta) \, F^*\left(\alpha + \frac{\xi}{\lambda}, \beta + \frac{\eta}{\lambda}\right) d\alpha \, d\beta$$
$$= F \star F^*$$
$$= \langle F_1 F_2^* \rangle_{\text{spatial}}.$$

Taking the time average of both sides, and interchanging the order of spatial integration and time averaging, we have

Fourier transform of $\langle f f^* \rangle \propto$ spatial average of $\langle F_1 F_2^* \rangle$.

Since time averages such as $\langle F_1 F_2^* \rangle$ are independent of spatial position in the Fraunhofer region, it follows that

$$\overline{T} \propto \langle F_1 F_2^* \rangle.$$

Hence an interferometer measurement at a single spacing, which is known from Sect. 55 to be a measurement of a single value of $\overline{T}$, can also be regarded as a measurement of the time correlation $\langle F_1 F_2^* \rangle$ of the field phasors F_1 and F_2 produced at the two aerials by the celestial distribution T.

To show this directly, we let the distribution be *slowly* in motion as indicated by a distribution function $T(x - Vt, y)$. Then the field produced at one aerial is $F_1 e^{i \frac{1}{2} \varphi}$, where

$$\varphi = 2\pi S \sin Vt \approx 2\pi S Vt$$

and the received voltage at the terminals of the aerial pair will be

$$V = V_1 + V_2 = \alpha_1 e^{-i\delta_1} F_1 e^{-i\frac{1}{2}\varphi} + \alpha_2 e^{-i\delta_2} F_2 e^{i\frac{1}{2}\varphi},$$

where the factors $\alpha_1 e^{-i\delta_1}$ and $\alpha_2 e^{-i\delta_2}$ allow for the attenuation and phase delay in each arm of the transmission line. The instantaneous power will be proportional to $V V^*$ and we have

$$V V^* = \alpha_1^2 F_1 F_1^* + \alpha_2^2 F_2 F_2^* + \alpha_1 \alpha_2 [F_1 F_2^* e^{i(-\varphi - \delta_1 + \delta_2)} + F_1^* F_2 e^{-i(-\varphi - \delta_1 + \delta_2)}].$$

As time elapses the mean value $\langle V V^* \rangle$ of the instantaneous power will measure the available power at the terminals, and under conditions where

$$\langle F_1 F_1^* \rangle = \langle F_2 F_2^* \rangle,$$

i.e. when behavior at one point is, on the average, the same as that at another point nearby, we have

$$\langle V V^* \rangle = \langle V_1 V_1^* \rangle + \langle V_2 V_2^* \rangle +$$

$$+ \frac{\sqrt{\langle V_1 V_1^* \rangle \langle V_2 V_2^* \rangle}}{\langle F_1 F_1^* \rangle} [\langle F_1 F_2^* \rangle e^{i(-\varphi - \delta_1 + \delta_2)} + \langle F_1^* F_2 \rangle e^{-i(-\varphi - \delta_1 + \delta_2)}]$$

$$= \langle V_1 V_1^* \rangle + \langle V_2 V_2^* \rangle + 2|\Gamma| \sqrt{\langle V_1 V_1^* \rangle \langle V_2 V_2^* \rangle} \cos(-2\pi S Vt - \delta_1 + \delta_2 + \text{pha}\,\Gamma),$$

where

$$\Gamma = \frac{\langle F_1 F_2^* \rangle}{\langle F_1 F_1^* \rangle}$$

and

$$|\Gamma| = \frac{|\langle F_1 F_2^* \rangle|}{\langle F_1 F_1^* \rangle}.$$

The final expression shows that as time elapses the available power rises and falls sinusoidally above and below a mean level $\langle V_1 V_1^* \rangle + \langle V_2 V_2^* \rangle$ with a depth of modulation characterized by the coefficient $|\Gamma|$, a dimensionless parameter which can be determined from

$$|\Gamma| = \frac{\langle V V^* \rangle_{\text{max}} - \langle V V^* \rangle_{\text{min}}}{\langle V V^* \rangle_{\text{max}} + \langle V V^* \rangle_{\text{min}}} \cdot \frac{1}{2} \left[\sqrt{\frac{\langle V_1 V_1^* \rangle}{\langle V_2 V_2^* \rangle}} + \sqrt{\frac{\langle V_2 V_2^* \rangle}{\langle V_1 V_1^* \rangle}} \right].$$

The first factor on the right hand side is the exact analog of the "visibility" or "contrast" of optical interference fringes as introduced by MICHELSON. The quantity Γ, which is the normalized time correlation of the field phasors at the two aerials, is the same as the "complex degree of coherence" of ZERNIKE[1]. Where the two aerials have equal gains and equally efficient transmission lines, so that $\langle V_1 V_1^* \rangle = \langle V_2 V_2^* \rangle$, the modulus of Γ is equal to the visibility of the fringes. In the case of unequal available powers from each of the aerials separately,

[1] F. ZERNIKE: Physica, Haag 5, 785 (1938).

$|\Gamma|$ could be determined from an observation of fringe visibility when the proportions had been established by a further measurement. The phase of Γ depends on an observation of fringe epoch and is possible though difficult.

We have thus shown that a simple interferometer observation measures the normalized time correlation of the field phasors at the aerials, and we shall occasionally borrow the terms "complex degree of coherence" and "visibility", as already with the word "fringe". The visibility, which is a parameter principally of the electromagnetic field, depends also on inequality of the aerial gains and feeder losses.

The complex coherence appears to play a fundamental role in optics since it is an observable quantity (unlike the field vectors themselves). It has interesting properties[1] one of which is to remain unchanged on planes parallel to a radiating aperture, neglecting evanescent fields. We have already established the property that the complex coherence is the Fourier transform of the normalized brightness temperature distribution over the source.

57. Effect of bandwidth. When an interferometer consisting of two infinitesimal elements at a fixed spacing of p nominal wavelengths forms part of a system

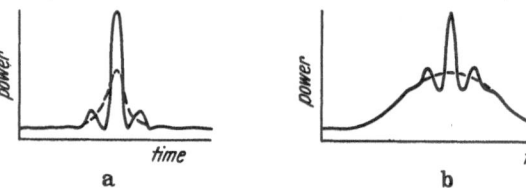

a

b

Fig. 44 a and b. Power received by the interferometer of Fig. 42 from a point source when the resolution is increased (a) by widening the aerials, (b) by widening the wavelength band.

sensitive to a band of wavelengths surrounding the nominal wavelength, then it can respond to a band of spatial frequencies surrounding $s = p$. If the spacing measured $p - q$ wavelengths at the longest wavelength and $p + q$ at the shortest, then the effect would resemble the result of making each element q wavelengths wide while keeping to monochromatic operation. In each case the effect is to extend the range of sensitivity to spatial frequency components, but with the difference that widening the elements widens the response around $s = 0$ whereas broadening the wavelength band does not. Consequently the responses to a point source, though sharpened in each case, reveal differences. In Fig. 44 we see curves corresponding to that of Fig. 42 when conditions are modified so that in (a) the width of the elements is greatly increased and in (b) the band of wavelengths is opened out. The broken lines show what would be received on a single aerial. Case (a) is analogous to the monochromatic Fraunhofer diffraction pattern of a pair of slits and case (b) is connected with the white and colored fringes observed when two (narrower) slits are illuminated by white light from a point source.

It is rewarding to contemplate the Fourier transforms of curves (a) and (b), but there is a complication to be borne in mind when extended wavelength bands are considered, namely the wavelength spectrum of the aforesaid "point source". Changing to the domain of angles we can say in general that the response of a broad band system is given by

$$\int_0^\infty D_\lambda(\vartheta, \varphi)\, S_1(\lambda)\, S_2(\lambda)\, d\lambda,$$

[1] J. A. Ratcliffe: Rep. Progr. Phys. **19**, 188 (1956). — E. Wolf: Proc. Roy. Soc. Lond., Ser. A **225**, 96 (1954); **230**, 246 (1955). — A. Blanc-Lapierre and P. Dumontet: Rev. Opt. (theor. instrum.) **34**, 1 (1955).

where D_λ is the directivity at wavelength λ, and $S_1(\lambda)$ and $S_2(\lambda)$ are suitably normalized functions describing the source emission spectrum and the spectral sensitivity of the system.

Restricting attention to a system which is sensitive over a narrow band $\Delta\lambda$ and operates at a nominal wavelength $N\,\Delta\lambda$ we see that each wavelength in the band $\Delta\lambda$ produces a pattern such as that of Fig. 42, the central fringes coinciding but the outer fringes tending to cancel. The extent of the beating pattern will be of the order of N distinct fringes since the N-th fringe of wavelength $\lambda-\tfrac{1}{2}\Delta\lambda$ falls just between the N-th and $(N-1)$-th fringes of the nominal wavelength λ.

The precise way in which the obliteration of the fringes occurs depends on the function $S_1(\lambda)\,S_2(\lambda)$. The theory of this phenomenon is precisely that applicable to MICHELSON's attempts to determine the profile of spectral lines from the visibility of the fringes produced by two interfering beams.

58. The sea interferometer. The first application of radio interferometry to celestial objects was made by McCREADY, PAWSEY and PAYNE-SCOTT[1] who determined the angular size, and position on the Sun's disc, of those discrete sources of radio energy localized in the corona over sunspot groups, which are now referred to as noise storms. They did this with a radar aerial whose original purpose was for height finding of aircraft by means of an accurate range measurement combined with an accurate measurement of angle of elevation.

The requisite precision in angle of elevation was achieved by placing a single aerial on a high cliff overlooking the sea and allowing rays reflected from the sea surface to interfere with direct rays. In the application to height finding, ambiguities due to the fineness of the lobe structure were resolved with the aid of another aerial placed above the first. If one replaces the highly reflecting sea surface by a second aerial at the image of the first, one has a two-element interferometer and so it is not necessary to repeat the theory given by McCREADY, PAWSEY and PAYNE-SCOTT.

Some differences only need be noted. Because of a phase reversal on reflection the first fringe falls above the locus of zero path difference as in LLOYD's optical arrangement for producing interference at glancing incidence from a mirror. In actual fact the first fringe was a degree below the true horizon, mainly because of tropospheric refraction. The interference pattern has a sharp beginning and consequently permits higher angular resolution than the corresponding two element device; for example, otherwise unresolvable sources with different times of rising might be distinguished.

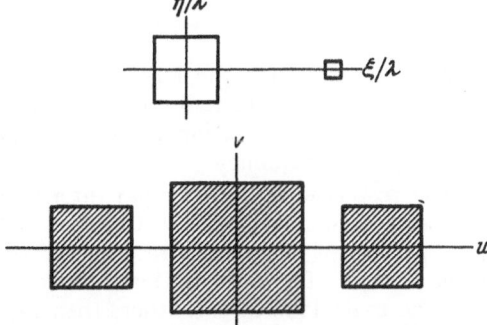

Fig. 45. Aperture distribution representing a large aerial and satellite (above) and the area of sensitivity in the uv-plane (below).

59. Unequal interferometer. A very large aerial, not readily movable may be accompanied by a smaller satellite aerial which combines with it to form an interferometer. This arrangement is becoming important and affords an opportunity to discuss a simple case in two dimensions. Fig. 45 shows a large square aerial supplemented by a small

---

[1] L. L. McCREADY, J. L. PAWSEY and RUBY PAYNE-SCOTT: Proc. Roy. Soc. Lond., Ser. A, **190**, 357, (1947).

square one represented by outlines in the $\xi\eta$-plane. The outlines in the uv-plane within which $\bar{A}$ is not zero are readily ascertainable graphically by copying the upper pattern on a sheet of transparent paper and sliding it about so that the copy is externally tangent to the original. A point moving with the copy then traces out the lower diagram.

Full evaluation of $E \star E^*$ requires a statement of the aperture distributions; for example, if they are uniform, the central island on the uv-plane is pyramidoidal with a central peak, and the adjacent islands are flat plateaux with steeply sloping borders.

The general conclusion is that the arrangement has the sensitivities of the separate aerials and in addition is sensitive to a set of spatial frequencies surrounding the value corresponding to the spacing between centers of the aerials. The extent of this latter set is considerable, being a little more than that corresponding to the dimensions of the large aerial.

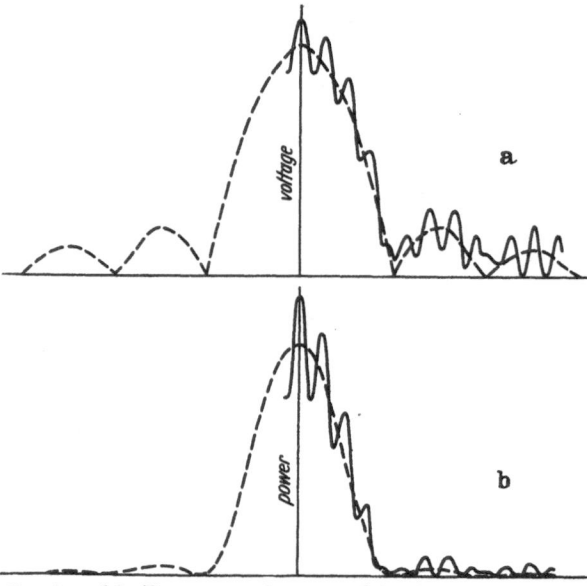

Fig. 46 a and b. The response of an unequal interferometer to a point source: (a) voltage, (b) power.

Since interest may reside principally in the flanking islands, not in the central one, an advantage may accrue from reduction in sensitivity in the central island. The switched interferometer and the method of post-detection correlation, both discussed below, would have this effect. It might appear that attenuation of the signal from the large aerial would help, a procedure which can readily be studied by reducing the aperture function for the large aerial before taking the autocorrelation of the composite aperture distribution. An interesting optimum sensitivity property exists, but the attenuation reduces the power received from the components of higher spatial frequency.

The response of the unequal interferometer to a point source is necessarily complicated but the cross section parallel to the ξ-direction is simple. Fig. 46 gives in broken outline the response of the large aerial in terms of the modulus of the voltage. The contribution from the small aerial, which is relatively constant over the main lobe of the other, then rapidly cycles through relative phases to produce the undulation of constant amplitude, shown as a full line (above). The power response (below) obtained by squaring, then shows how the high frequency band has been broadened and strengthened.

60. Payne-Scott and Little's interferometer. In Sect. 55 we described how the envelope of the radiation pattern of an interferometer could be steered in direction by rotating each aerial on an axis through itself; but the direction of the individual fringes, which is determined by the base line and the wavelength, does not change.

Now if the cables connecting the two aerials together are not of precisely equal electrical length, the central fringe will be displaced from the median plane by such an angle that the excess path from a distant source to the aerial with the shorter cable just compensates the cable defect. The electrical length of a cable can readily be controlled by changing its physical length or its phase velocity and a device for doing this may be called a phase-shifter. A phase shift changing progressively with time will cause the fringes to sweep across the sky, and to grow and decay in such a way that their envelope remains fixed.

An alternative explanation of this system can be given in terms of the interference between two waves of slightly differing frequency, since a steadily changing phase means a change in frequency.

Since the aerial pattern as a whole is not swept over the sky, the received power cannot be expressed as a convolution integral and the simple aerial smoothing theory does not apply. This was also the case with the pair of aerials on equatorial mountings and it was connected with the non-rigidity of the arrangement, an interpretation which can also be placed on the present case.

Let the radiation pattern of a single aerial be $A(x, y)$, let the true distribution of brightness temperature be $T(x, y)$, and let $T_a(x, y)$ be the distribution which would be observed with a single aerial. Then if the aerials are placed S wavelengths apart in the ξ-direction and a phase shifter in one arm introduces phase at the rate of Ω radians per second, the radiation pattern at time t is

$$A(x, y) \left[1 + \cos \left(2\pi S x - \Omega t\right)\right]$$

and the received power at time t when the aerial is pointed at (x, y) is given by

$$\iint T(x' + x, y' + y) A(x', y') \left[1 + \cos \left(2\pi S x' - \Omega t\right)\right] dx' dy'$$
$$= T_a(x, y) + \cos \Omega t \iint \cos 2\pi S x' \, T(x' + x, y' + y) A(x', y') \, dx' dy' +$$
$$+ \sin \Omega t \iint \sin 2\pi S x' \, T(x' + x, y' + y) A(x', y') \, dx' dy'.$$

The power received with the aerials pointed in a fixed direction, say $(0, 0)$, is

$$T_a(0, 0) + M \cos (\Omega t - \alpha),$$

where

$$M e^{i\alpha} = \iint e^{i 2\pi S x'} T(x', y') A(x', y') \, dx' dy' = \overline{TA}(S, 0).$$

Thus as time elapses, the power received with fixed aerials consists of a steady component equal to what would have been received on a single aerial plus a periodic component with period $2\pi/\Omega$. From the amplitude and phase of the periodic component one gets one Fourier component of $T(x, y) A(x, y)$, namely $\overline{TA}(S, 0)$. Since the steady component $T_a(0, 0)$ can be expressed as $\overline{TA}(0, 0)$, it follows that PAYNE-SCOTT and LITTLE's device is precisely equivalent to a simple two-element interferometer in the limiting case of small elements for which $A \rightarrow$ const; it merely obtains the same information more quickly, and that is the purpose for which it was devised[1].

By increasing the bandwidth of the system one obtains greater sensitivity but at the same time the fringes of high order die out as described in Sect. 57. An advantage of narrow bandwidth in this instrument is the ability to work over a large area of sky without adjustment of the aerials. Increased bandwidth brings with it the need for a phase delay in one arm which will cause the zero-order fringe to follow the Sun; this is in addition to the physical following necessitated by the limited beamwidth of the aerials themselves.

---

[1] See Austral. J. Sci. Res. A **4**, 489 (1951); The Observatory **70**, 185 (1950).

Payne-Scott and Little introduced preamplifiers into their system at each aerial for the purpose of counteracting signal loss in the cables. This would have been impossible in a simple interferometer without exceptional equalization and stabilization of gains, phase shifts and noise figures, but in the phase-sweeping system, and in the phase-switching system discussed in the following section, moderate instability and inequality affect only the "steady" component, not the periodic component.

61. Ryle's interferometer. Of the two bands of Fourier components received by an interferometer, only the high frequency band contains information that

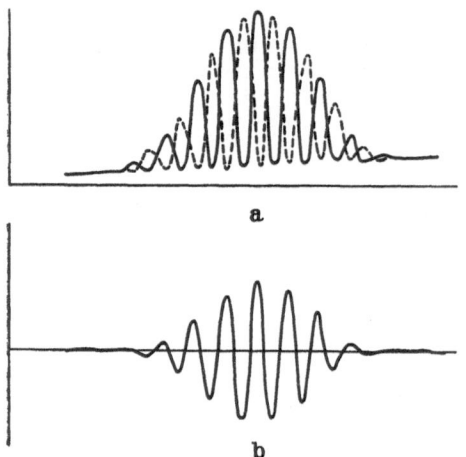

a

b

Fig. 47 a and b. (a) The two separate responses of a phase-switched interferometer. (b) The response of the system to a point source.

could not have been obtained with a single aerial and so it would not be a loss if the low frequency band were filtered out. Such action would also bring advantages, for the steady background shown in Fig. 42 contains a contribution of galactic and extragalactic origin which rises and falls in strength as the day elapses and is embarrassing when large compared with the record of a desired faint source; this would be removed by filtering. So also would slow changes in receiver noise.

A method of doing this was introduced by Ryle[1]. By switching an extra half wavelength of cable in and out of one arm of the interferometer at a rapid rate such as 25 Hz, the fringe pattern can be switched back and forth, under its envelope, in a discontinuous version of the phase-sweeping scheme. Ryle describes it as follows.

When the interferometer is in one of its two possible conditions, the response to a point source is

$$A(x)\left[1 + \cos 2\pi S x\right]$$

as shown by the heavy line in Fig. 47 (a), and when in the other, the response, as shown by the broken line, is

$$A(x)\left[1 - \cos 2\pi S x\right].$$

When the interferometer alternates rapidly between its two conditions the received power has a steady component and a rapidly alternating component of amplitude

$$2A(x)\cos 2\pi S x$$

which may be rectified in a phase sensitive detector and recorded. The resulting record of Fig. 47 (b) is in effect the difference between the two upper curves.

62. The Mills cross. Two aerials A_1 and A_2 (Fig. 48), one greatly extended in the ξ-direction, the other in the η-direction, form the two elements of an interferometer. The first receives from a narrow strip of sky a_1 and the second from a similar strip a_2 at right angles (not shown). When the two are connected together in phase their reception pattern is as shown at b, point sources in either a_1 or a_2 being received, and sources in both a_1 and a_2 being received more strongly.

[1] M. Ryle: Proc. Roy. Soc. Lond., Ser. A **211**, 351 (1952).

When the two aerials are connected together in phase opposition, which may be arranged by inserting a half-wavelength of cable into one arm, sources not in both a_1 and a_2 are received as before, but sources lying in the intersection of a_1 and a_2 are not received since their contributions to A_1 and A_2 are equal and combined in opposition [Fig. 48(c)]. Now if the system is caused to alternate rapidly between its two possible states, then the received power will comprise a steady part due to sources in only one beam and an alternating part due to sources lying in the intersection. The steady part may be ignored and the strength of the alternating component recorded by the use of a phase-sensitive detector.

Such a system was first constructed by MILLS and LITTLE[1] who thus achieved the angular resolution corresponding to a conventional aerial occupying the broken square outline, which would receive from an area of sky of size d. A photograph of the 1500 foot Mills cross for meridian use is shown in Fig. 49a. Fig. 49b

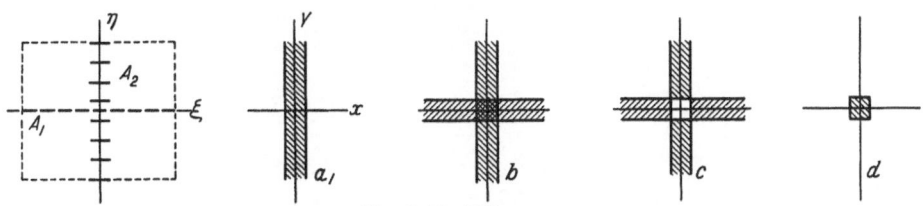

Fig. 48. The Mills cross.

shows an application [2] of the cross principle employing fully steerable elements, while Fig. 49c illustrates the quality of resolution obtainable on the sun.

The beamwidth of such a system is much less than would be achieved by a conventional aerial having the same collecting area, or same number of elements, and apparently violates the theoretical relation between the area and beamwidth of an aerial. However, while the system contains two aerials, it is not an aerial itself.

If the patterns represented schematically by a_1 and a_2 are written as $V_1(x, y)$ and $V_2(x, y)$, where V_1 and V_2 are respectively the phasors representing the voltages produced in the two separate aerials by a point source at (x, y), then the available power in the first condition of combination is proportional to

$$[V_1(x, y) + V_2(x, y)] \times [V_1(x, y) + V_2(x, y)]^*$$

and in the second condition,

$$[V_1(x, y) - V_2(x, y)] \times [V_1(x, y) - V_2(x, y)]^*.$$

The difference between these two expressions, which represents the amplitude of the alternating component, is proportional to

$$V_1(x, y)V_2^*(x, y) + V_2(x, y) V_1^*(x, y).$$

The power response of the system to a point source is thus the scalar product of the voltage responses of the separate aerials.

Such a pattern has sidelobes which are stronger relative to the main beam than is the case with the patterns of conventional aerials. In surveying the sky for discrete sources, there is a possibility that a strong source in the direction of a sidelobe may be mistaken for a faint source in the main beam. For this reason MILLS tapered his arrays to smooth out the lobe structure. The most

[1] B.Y. MILLS and A.G. LITTLE: Austral. J. Phys. **6**, 272 (1953).
[2] R. N. BRACEWELL: Inst. Radio Engrs. National Convention Record **5**, part I, 68 (1957).

suitable degree of taper in a given application will depend on a balance between (i) spurious information on the one hand and (ii) lost information due to failure of the widened beam to resolve and to obliteration by the increased general side radiation on the other.

Arrangements for steering the beam on the meridian by introducing progressive phase shifts in a north- south dipole array have proved feasible in practice

b

Fig. 49 a, b. (Photo.) (a) The 1500 foot Mills cross at Fleurs, New South Wales, wavelength 3.5 m, beamwidth 50 minutes of arc, (b) a cross of equatorial paraboloids at Stanford, California, wavelength 9 cm, beamwidth 3.1 minutes of arc.

and it is clear that the Mills cross represents a great instrumental advance in those fields where resolution, not sensitivity, is the desideratum.

The cross principle lends itself to the use of aerials of very great physical extent and, because of this, substantial effective area accrues even when aerials A_1 und A_2 are only about a wavelength wide. For future astronomical observations requiring both high resolution and high sensitivity the aerials would also have to be wide. Two distinct possibilities follow, each of which allows construction in sizes much larger than the largest possible steerable giant paraboloid. For a meridian instrument, a wide east-west aerial could conveniently be tilted and could assume the form of a parabolic cylinder with focal-line feed (as in Fig. 30). A wide north-south aerial could be a fixed parabolic cylinder steered by introducing progressive phase shifts along the focal line; alternatively the north-south aerial could be an array of more or less closely spaced tiltable segments. A second possibility, already being exploited both at Sydney and Stanford, is to split each aerial into an array of equatorially mounted segments (Sect. 64), thus relieving the restriction to transit observations (Figs. 35c and 49b).

Whilst the theory given above is quite valid there is a subtlety that may be overlooked. If the aerials A_1 and A_2 are represented by aperture distributions

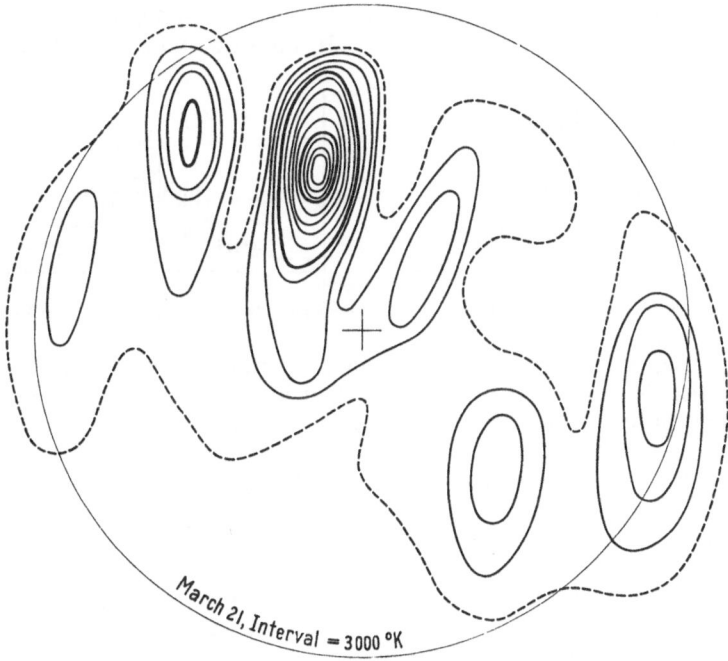

Fig. 49c. Picture of the microwave sun showing the quality of resolution obtainable at 9 cm.

they may not overlap since the energy falling at $\xi = 0$, $\eta = 0$ can only be absorbed once. It is therefore necessary that one or both aerials have a defect at the origin, let us say A_1. Then $V_1(x, y)$ comprises the narrow strip a_1 and also a shallow negative part distributed in accordance with the voltage reception pattern of the missing part. The response of the system to a point source then has negative parts which bring its integral to zero; which is necessary for a system so designed that it cannot respond to a uniform distribution. In Sect. 64 we refer to this matter again.

63. The Brown and Twiss system. As we have seen, the measurement obtained with a pair of spaced aerials yields a value of $\overline{T}$, normalized in the case of a simple interferometer, otherwise not. For a very small source $\overline{T}$ falls only slowly from its value at the origin, and so very great aerial spacings are required to reveal the appreciable change in $\overline{T}$ that will yield a size measurement. The large spacing in turn requires the preamplification which is only feasible with a phase-switched interferometer, and then careful absolute measurements at at least two spacings are necessary since a single phase-switched interferometer record does not by itself yield a value of fringe visibility. For these reasons difficulty was encountered in the first attempts to measure the diameter of the source in Cygnus, which though strong proved to have an extremely small angular diameter, of the order of one minute of arc. From the inverse theorem of Sect. 54 it follows that independent observations of $\overline{T}$ cannot be made at intervals less than about 3000 wavelengths. This places a considerable strain on technique and was faced in three ways. Smith[1] obtained measurements to 423 wavelengths ($\lambda = 1.4$ m) giving very great care to equalizing and stabilizing cable loss and preamplifier gains. Mills[2] substituted a radio link for the cable and obtained measurements out to 3000 wavelengths ($\lambda = 3$ m), taking particular care to preserve phase in an ingenious fashion. It appears, however, that it is not

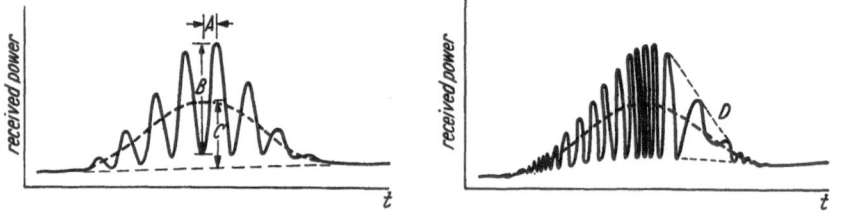

Fig. 50. Simple interferometer records of a discrete source in which phase is preserved (left) and allowed to drift (right).

necessary to preserve phase if only one shape parameter, viz. a second moment of T, is all that is sought. And it is clear that with the moderate spacings mentioned above only one parameter would be justified. Brown, Jennison and Das Gupta[3] based their measurement on this fact.

Consider a simple interferometer with which a single record at a large spacing has been obtained on a very small source, and from this record suppose that a measurement of complex degree of coherence Γ has been taken. Then $|\Gamma|$ yields the size of the source and pha Γ relates the center of gravity of the source to the median plane of the interferometer. It will now be shown that a record taken without care to preserve phase yields $|\Gamma|$ the desired quantity but loses pha Γ a quantity which in any case is normally considered difficult to measure. In Fig. 50 is shown on the left a record from the simple interferometer, which should be compared with the record of a point source shown in Fig. 42. From the depth of the minima it is apparent that the spacing is sufficiently great to reveal a fringe visibility $B/2C$ appreciably less than unity, which can be used as explained in Sect. 68 to determine the source width. The distance A, taken in conjunction with the fringe period and corrections for collimation gives pha Γ. On the right hand side of Fig. 50 we see the effect of an irregular phase drift. Clearly the envelopes from which fringe visibility are determined are not lost, except momentarily by an accident such as that indicated at D.

On this basic point regarding phase Hanbury Brown based a new system which is described by Brown and Twiss[4] and by Jennison and Das Gupta[5]. The independence of phase was to permit the extension of aerial spacings to 50 km

[1] F. G. Smith: Proc. Phys. Soc. Lond. B **65**, 971 (1952).
[2] B. Y. Mills: Nature, Lond. **170**, 1063 (1952).
[3] R. H. Brown, R. C. Jennison and B. K. Das Gupta: Nature, Lond. **170**, 1061 (1952).
[4] R. H. Brown and R. Q. Twiss: Phil. Mag., Ser. VII **45**, 663 (1954).
[5] R. C. Jennison and M. K. Das Gupta: Phil. Mag., Ser. VIII **1**, 55 (1956).

which would be necessary if, as was thought, the radio sources should prove to be of stellar dimensions. The system actually built incorporates other, inessential, departures from previous practice, viz. (i) the two received signals were detected before being combined, (ii) the multiplication effected in the phase-switched system by differencing the squares of the sum and difference of two voltages, was carried out by a special envelope multiplying circuit referred to as a correlator, (iii) the time-varying quantities whose cross-correlation was evaluated were the squares of the envelopes of the received signals.

For a full understanding of the Brown and Twiss system it is therefore necessary to examine the relationship between $\langle F_1 F_2^* \rangle$, the quantity involved in an observation of fringe visibility with a simple interferometer (Sect. 56), and $\langle F_1(t) F_1^*(t) \cdot F_2(t+\tau) F_2^*(t+\tau) \rangle$ the cross correlation between the squares of the envelopes of $F_1(t) e^{i\omega t}$ and $F_2(t) e^{i\omega t}$. In practice τ will be arranged to be zero by observing in the median plane, or, if the source under study is off to the side, by inserting a compensating time delay. It can be shown (BRACEWELL[1]) that

$$\frac{\langle (F_1 F_1^* - \langle F_1 F_1^* \rangle)(F_2 F_2^* - \langle F_2 F_2^* \rangle) \rangle}{\langle (F_1 F_1^* - \langle F_1 F_1^* \rangle)^2 \rangle} = \left[\frac{\langle F_1 F_2^* \rangle}{\langle F_1 F_1^* \rangle} \right]^2 = |\Gamma|^2$$

i.e. the fluctuating parts of the squared envelopes $F_1 F_1^*$ and $F_2 F_2^*$ are correlated; but not as strongly as are the phasors F_1 and F_2^*, since the first correlation coefficient is the square of the second.

Receiver noise must not be neglected in this discussion. When visibility is being measured from a record taken with a simple interferometer the presence of receiver noise may hinder the measurement but does not change its character. However, the correlation between the squared envelopes of two oscillations containing steady receiver noise components large compared with the contributions due to the fields F_1 and F_2 is not the same as in the absence of noise and proves to be equal to the correlation between the field phasors. The important point is that the correlation is not destroyed by demodulating the radio frequency signals. The essence of the proof for the noise-free case, has been concisely set out by WOLF[2] who shows that the correlation between the fluctuating parts of the instantaneous field intensities is equal to twice the square of the correlation between the instantaneous real fields at two points.

b) Multi-element interferometers.

64. CHRISTIANSEN's interferometer. Consider a long array of identical aerials, such as paraboloids, spaced at equal intervals of B wavelengths along the ξ-axis. The aperture distribution may be expressed as the convolution of the distribution over one aerial with the function

$$\sum_{n=1, 3, 5 \ldots}^{M-1} 2\delta \left(\frac{\xi}{\lambda} \pm \frac{nB}{2}, \frac{\eta}{\lambda} \right),$$

which is a row of M impulses in the $\xi\eta$-plane. Consequently the field radiation pattern is the product of the pattern for a single aerial with the Fourier transform of the above set of impulses. By taking the impulses in symmetrical pairs, the required transform can be expressed as a sum of $\frac{1}{2}M$ cosine functions

$$2 \sum_{n=1, 3, 5 \ldots}^{M-1} \cos \pi n \, Bl,$$

---

[1] R. N. BRACEWELL: Proc. Inst. Radio Engrs. **46**, 97 (1958).
[2] E. WOLF: Phil. Mag., Ser. VIII **2**, 351 (1957).

which is the Fourier series for a periodic function consisting of overlapping functions of the form

$$\frac{\sin [M \pi B l]}{\pi B l}$$

repeated with alternation of sign at intervals $1/B$. This latter expression will be recognized as the field pattern of a uniform array $M B$ wavelengths long and of

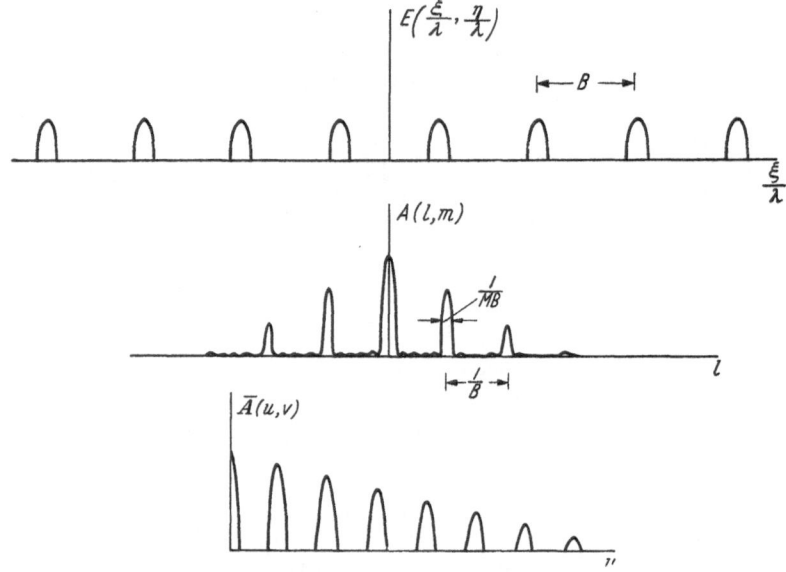

Fig. 51. The aperture distribution, radiation pattern, and spectral sensitivity function of a Christiansen array.

zero width. The power radiation pattern, which we obtain by squaring the field pattern in this case, thus consists, when $M B$ is large, of a set of parallel sharp fringes whose profile is of the form

$$\left[\frac{\sin \pi M B l}{\pi l}\right]^2 ,$$

the whole multiplied by the radiation pattern of a single aerial.

These conclusions are illustrated in Fig. 51 which also shows the spectral sensitivity function.

This aerial array is analogous to a diffraction grating with a total of M lines and an aperture of $M B$ wavelengths, in contact with a slit set perpendicular to the ruling. The repeated fringes correspond to the beams of various order, and the fringe profile corresponds to the diffraction pattern of the slit in the long direction. The envelope of the fringe pattern corresponds to the diffraction pattern of a single transparent element of the ruling, and for the closest analogy the width of the slit would be equal to the width of a single transparent element. The first array of this kind was built by Christiansen and Warburton[1] and had 32 elements each spaced approximately 30 wavelengths apart.

When a discrete source, such as the Sun, whose angular extent is less than the fringe spacing passes through the fringe system, a record is obtained which is virtually identical with what would be obtained with a uniform aperture a little longer than the array. The omission of aperture excitation between the

[1] W.N. Christiansen and J.A. Warburton: Austral. J. Phys. **6**, 262 (1953).

elements thus results in no loss of resolution, though there is a loss in available power. This is another example of the inverse discrete interval theorem. The spectral sensitivity function shows that the spatial spectrum will be sampled at uniform intervals and nothing will be lost thereby if the object does not extend beyond a certain finite width. In the case of the Sun, which is approximately 0.01 radians in diameter, the gap between the aerials could not exceed 100 wavelengths, and should be less to allow for the fringe width.

The importance of a new idea of the present kind is that a great advance over previous technique becomes possible. Furthermore the idea is transferable to other systems which combine aerials, such as the Mills cross, and a future application to arrays of giant aerials of the size just now coming into existence can be foreseen.

65. Multi-element phase-switched systems. The use of phase-switching in conjunction with two general aerial arrays had been contemplated by RYLE[1] and various complicated systems are conceivable. One interesting example developed by COVINGTON and BROTEN[2] resulted in the achievement of twice the angular resolution which was believed possible at the time within the overall dimensions.

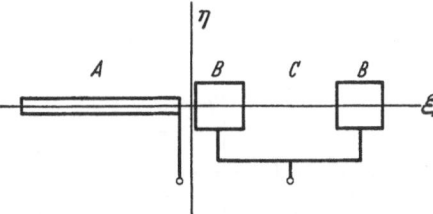

Fig. 52. A compound phase-switched system.

A long narrow existing aerial A (Fig. 52) was combined with a two-element interferometer BB by phase switching. The (power) response to a point source can be shown to be the product of the (voltage) responses of (i) the long array, (ii) an interferometer consisting of two isotropic aerials at BB, (iii) an interferometer of spacing AC, (iv) a single aerial B. The net result was a beam which in one dimension was about half that for an ordinary aerial of the same overall length in the ξ-direction.

Consider the general case of an aperture E_1 of any kind centered at $\xi = -\frac{1}{2}a$, $\eta = 0$ and another E_2 at $\xi = +\frac{1}{2}a$, $\eta = 0$, and let them be connected together through a phase-switch. Let

$$E_2 \star E_1^* = E_3$$

and

$$E_1 \star E_2^* = E_4.$$

Then the spectral visibility functions of the two states of connection are

$$\left[E_1\left(\frac{\xi + \frac{1}{2}a}{\lambda}, 0 \right) \pm E_2\left(\frac{\xi - \frac{1}{2}a}{\lambda}, 0 \right) \right] \star [\text{conjugate}]$$

$$= E_1\left(\frac{\xi + \frac{1}{2}a}{\lambda}, 0 \right) \star E_1^*\left(\frac{\xi + \frac{1}{2}a}{\lambda}, 0 \right) + E_2\left(\frac{\xi - \frac{1}{2}a}{\lambda}, 0 \right) \star E_2^*\left(\frac{\xi - \frac{1}{2}a}{\lambda}, 0 \right) \pm$$

$$\pm E_2\left(\frac{\xi - \frac{1}{2}a}{\lambda}, 0 \right) \star E_1^*\left(\frac{\xi + \frac{1}{2}a}{\lambda}, 0 \right) \pm E_1\left(\frac{\xi + \frac{1}{2}a}{\lambda}, 0 \right) \star E_2^*\left(\frac{\xi - \frac{1}{2}a}{\lambda}, 0 \right).$$

Consideration of the alternating part ($\pm$ signs), viz.

$$E_3\left(\frac{\xi + a}{\lambda}, 0 \right) + E_4\left(\frac{\xi - a}{\lambda}, 0 \right),$$

[1] M. RYLE: Proc. Roy. Soc. Lond., Ser. A **211**, 351 (1952).
[2] A. E. COVINGTON and N. W. BROTEN: IRE Transactions on Antennas and Propagation **AP—5**, 247 (1957).

will give the Fourier transform of the response of the system to a point source, subject to normalizing. When E_1 and E_2 are real and even functions of ξ, both E_3 and E_4 are equal to $P_1 P_2$, the product of the field patterns; hence the response of the system $\boldsymbol{A}(\vartheta, \varphi)$ is given by

$$A(\vartheta, \varphi) = P_1 P_2 \cos 2\pi a \, \vartheta.$$

This result applies to the example quoted above, to the Mills cross, and to the phase-switched interferometer with identical aerials. The more general result, found by substituting for E_3 and E_4 and transforming, is

$$(A - \vartheta, -\varphi) = P_2 P_1^* \, e^{i 2\pi a \vartheta} + P_1 P_2^* \, e^{-i 2\pi a \vartheta}$$
$$= |P_1 P_2^*| \cos \left[2\pi a \, (\vartheta - \vartheta_1)\right].$$

Interesting special cases include (i) the long uniform linear array phase-switched against a single aerial at one or at each end, which gives a flat spectral

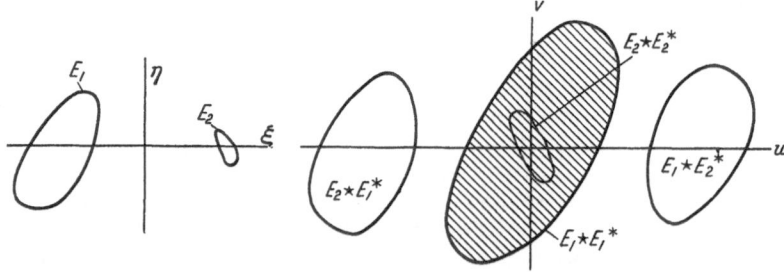

Fig. 53. Spectral sensitivity islands in the uv-plane for a phase-switched system comprising apertures E_1 and E_2. The shaded island, shown for completeness, is suppressed.

sensitivity function and doubles the resolving power of the array even if the single aerial has little directivity; (ii) two long perpendicular uniform arrays combined in a tee instead of a cross, thus doubling the resolution in one direction; (iii) a phase-switched unequal system which, referring to Fig. 45, would have the effect of suppressing the central island; (iv) a four element system comprising a pair of crossed variable-spacing two-element interferometers; (v) a long array with a single aerial at variable positions on the perpendicular bisector· In thinking about devices of this kind the two-dimensional spectral sensitivity functions are easier to obtain than the spatial responses, and are often more revealing. Fig. 53 illustrates a spectral sensitivity island diagram which was very readily drawn by sliding a copy of the $\xi\eta$ outline over itself as explained above. The shaded island is the one whose sign does not change when the sign of E_2 is reversed and which is therefore suppressed.

66. The 1100101 array. The spectral sensitivity functions so far illustrated exhibit bands (Fig. 43) or taper away towards their cut-off (Figs. 39, 51). The question may be asked whether an aperture distribution exists such that the spectral sensitivity function is constant out to its cut-off, i.e., can we have $\overline{A}(s) = \Pi\left(\dfrac{s}{2 s_c}\right)$, without resorting to phase-switching. This would imply that $A(x) = \dfrac{\sin 2\pi s_c}{\pi x}$ which is impossible since $A(x)$ cannot go negative. However $\dfrac{\sin 2\pi s_c x}{\pi x} + \dfrac{4 s_c}{3\pi}$ is non-negative and has a transform $\Pi\left(\dfrac{s}{2 s_c}\right) + \dfrac{4 s_c}{3\pi}\, \delta(s)$ which is an interesting possibility. In terms of equally spaced discrete elements of amplitude

0 or 1 we find the following arrays where the sequence of numbers $a_0\, a_1 \ldots a_n$

$E\left(\frac{\xi}{\lambda}\right)$	$\bar{A}(s)$
11	21
1101	3111
1100101	4111111

stands for $\sum\limits_{i=0}^{n} a_i\, \delta\left(\frac{\xi}{\lambda}-i\right)$ in the left hand column and $\sum\limits_{0}^{n} a_i\, \delta\,(s-i)$ in the right hand column.

The last entry in the table represents the 1100101 array, an array of four equal elements so spaced that there is one and only one pair of elements for each spacing from one unit up to six units, the full extent of the array. The spectral sensitivity function is therefore flat. However zero spacing occurs four times and so the array is unduly sensitive to the $s=0$ component, which is necessitated by the condition that $\bar{A}(s) \not< 0$. An array of parabolas spaced in this way has been built and demonstrated by Arsac[1]. There are no further arrays with the property shown by the three tabulated above, and in two dimensions no corresponding property has been found.

It is clear that the economy and simplicity of the 1100101 array will fit it for consideration when giant aerials become sufficiently numerous to group into arrays.

VI. Observing procedures and analysis of observations.

a) Discrete Sources.

67. Position of a point source. If one observes with a pencil beam, the position of a point source is simply the direction of the beam axis when maximum power is being received. An accompanying calibration is then necessary to determine the beam axis in terms of readings on the setting circles on the aerial mounting, the Sun being one convenient source for this purpose. A direct observing procedure is to fix the beam on the meridian and determine the right ascension from the time of transit, as revealed by the maximum in the record. Then take a series of drift curves at discrete intervals of declination equal to the peculiar interval of the aerial (Sect. 54). From the maxima of the drift curves determine the declination by interpolation, using if necessary the exact method of Sect. 70. The judgment of the observer will be necessary in removing any trend in the background and any confusing effects of nearby sources.

If one observes with an interferometer, and the great bulk of the early positional work was done in this way, various combinations of right ascension and declination will be measured and the final coordinates will be obtained by elimination, unless the interferometer is on an east-west baseline. In this case the right ascension is given by the time of transit, and the declination δ is given by

$$\cos \delta = \frac{86400}{2\pi S \tau},$$

where τ is the average period of the fringes in sidereal seconds, and the interferometer spacing is S wavelengths. The result follows from the factor $\cos\,(2\pi S V t - \beta)$ in the expression for the signal received from a discrete source (Sect. 55), which

---

[1] J. Arsac: C. R. Acad. Sci., Paris **240**, 942 (1955).

shows that the period of a fringe is $\dfrac{1}{SV}$, where V, the velocity of the source through the beam, is proportional to cos δ.

The order of accuracy of position determination has been minutely studied by MILLS and THOMAS[1] and by SMITH[2], the latter paper including results on the sea interferometer and many other arrangements. With great care absolute accuracy of one minute of arc was achieved in the location of the source in Cygnus, but only the most devoted observers will wish to resume this type of work since there are many reliably identified extragalactic sources whose optical positions may be used with confidence as points of reference for absolute positions within a few minutes of arc.

With the advent of aerials of great effective area serious confusion between adjacent sources has been encountered at the longer radio wavelengths. Let us assume that any continuous background is uniform and assess the limit to the number of resolvable sources set by random clustering of the sources themselves and the finite resolving power of the instrument. Since independent measures can be made only at the peculiar interval a survey of the whole sky will yield only a finite number of data equal to $4D/\zeta$, where D is the directivity and ζ the achievement factor of the aerial. Now if a saddle point in an observed distribution is taken as a criterion for resolution of two point sources, the distance between the two points must be at least 1.2 peculiar intervals if the sources are equal and more if they are not. Points scattered at random in two dimensions with a density n per unit area are separated on the average from their nearest neighbor by a distance $0.5\, n^{-\frac{1}{2}}$. Equating this distance to 1.2 peculiar intervals we find that the maximum number of resolvable sources in the whole sky is $\dfrac{D}{1.4\,\zeta}$, or approximately D. The number of sources per steradian is approximately $D/4\pi$ or one per effective solid angle Ω. In practice it is considered that there should be about 10 effective beamwidths per source in order to ensure that a substantial fraction of the apparerent sources are real, and perhaps 100 beamwidths per source if accurate flux density measurements are required.

68. Angular extent of a discrete source. The nulls in the response of an interferometer to a point source are filled in, if an extended source passes through the beam, by an amount depending on the width of the source. McCREADY, PAWSEY and PAYNE-SCOTT used this phenomenon to determine the width of the sources of solar noise situated above sunspot groups, and PAWSEY and BRACEWELL show that the width of a uniform one-dimensional source is given by

$$\frac{2\sqrt{3R}}{\pi S}\,,$$

where S is the spacing in wavelengths and R is the ratio of the minimum to the maximum of the interference pattern.

The observation is essentially one of fringe visibility at a single aerial spacing since the visibility as defined in Sect. 56 is equal to

$$\frac{1-R}{1+R}\,.$$

Consequently the data are equivalent to one Fourier component of the source distribution T, normalized with respect to $\bar{T}(0, 0)$. Provided the aerials are not

---

[1] B. Y. MILLS and A. B. THOMAS: Austral. J. Sci. Res. A **4**, 158 (1951).
[2] F. G. SMITH: Monthly Notices Roy. Astronom. Soc. London **112**, 497 (1952).

too widely spaced, this in turn is equivalent to a measure of the curvature of $\overline{T}$ at the origin, and by a theorem of Fourier transforms, this determines the standard deviation σ_x of T about its center of gravity, in the direction parallel to the baseline of the interferometer (taken as the x-direction). We find that

$$\sigma_x \approx \frac{\sqrt{R}}{\pi\,S}$$

for small R. It is felt that this is a more apposite statement of source size than $2W$ the width of the equivalent uniform strip, or D the diameter of the equivalent uniform disc. A fourth measure also in use is σ_r the standard deviation of the equivalent two-dimensional Gaussian source. These various quantities are related by

$$\sigma_x = \frac{2W}{2\sqrt{3}} = \frac{D}{4} = \frac{\sigma_r}{\sqrt{2}}\,.$$

If two width measurements at right angles were possible, $\sqrt{\sigma_x^2 + \sigma_y^2}$ would be a suitable size measure; but if only one is available and if the source possesses circular symmetry, or if circular symmetry forms a suitable basis, then σ_r or $\sqrt{2}\sigma_x$ is appropriate.

When R is small a single visibility observation thus measures the one-dimensional extent of a source; but when R is not small, the shape as well as the extent of the source are mixed together in the measurement.

69. Flux density of a discrete source. Suppose that a set of drift curves has been obtained at a set of properly spaced declinations and that they reveal a discrete source superimposed on a continuous background. The observer will often feel sufficiently confident to subtract the background by taking account of the surrounding continuum, which may appear to him to be relatively free of other discrete sources. Occasionally, helpful information on other frequencies will also be available. We shall assume that the background has been satisfactorily subtracted, leaving a remainder $T_a(x, y)$.

Now by Sect. 49 the flux density S of the source is given by

$$S = \frac{2k}{\lambda^2} \iint T\,dx\,dy,$$

where T is the true temperature distribution over the source. But we have only T_a. However

$$\iint T_a\,dx\,dy = \overline{T}_a|_{u=v=0}$$
$$= [\overline{A}\,\overline{T}]_{u=v=0}$$
$$= \overline{T}|_{u=v=0},$$

provided $A(x, y)$ is normalized so that

$$\iint A\,dx\,dy = 1,$$

whereupon

$$\overline{A}|_{u=v=0} = 1.$$

But

$$\iint T\,dx\,dy = \overline{T}|_{u=v=0},$$

therefore

$$S = \frac{2k}{\lambda^2} \iint T_a\,dx\,dy,$$

and the flux density is correctly calculated from a knowledge of T_a only.

We now show that the integral can be evaluated merely by summing discrete values. If $\overline{T}_a = 0$ for $|u| \geq \frac{1}{2}$ and for $|v| \geq \frac{1}{2}$, then

$$\iint T_a\, dx\, dy = 4\, [^2III(2u, 2v) * \overline{T}_a]_{u=v=0}$$

$$= [\overline{^2III(\tfrac{1}{2}\, x, \tfrac{1}{2}\, y)\ T_a}]_{u=v=0}$$

$$= \iint {}^2III\, (\tfrac{1}{2}\, x, \tfrac{1}{2}\, y)\ T_a\, dx\, dy$$

$$= 4 \sum_{\mu=-\infty}^{\infty} \sum_{\nu=-\infty}^{\infty} T_a\, (x - \alpha\mu,\ y - \beta\nu)$$

provided α and β are both less than 2.

The essence of this proof is that the integral of T_a equals the height at the origin of the island representing $\overline{T}_a$ in the uv-plane. The distribution $^2III\, T_a$ (which contains information about T_a only at the sampling points where $^2III \neq 0$) has a transform consisting of an array of islands identical with $\overline{T}_a$ but centered on the points of a rectangular lattice. As before, the integral of $^2III\, T_a$ is equal to the height of the island system at the origin; but this is the same as before, provided the islands do not crowd too closely. When the sampling interval is so coarse that the islands just close in and touch each other, we have sampling at the peculiar interval (Sect. 54). However, since we only require the height of the central island at the origin to remain unchanged, we may allow overlapping — just so much that the edges of the nearest islands do not reach the origin. This allows the sampling interval to be twice as coarse as the peculiar interval.

Thus we correctly evaluate S by summing just one in four of the values of T_a necessary to define T_a. This result permits a time saving in observation and reduction or alternatively permits certain cross-checks when redundant data are available.

In the foregoing, T_a was restricted, as in Sect. 50, to a loss free aerial. Hence effective aerial temperatures actually observed will be less by a factor η, the aerial efficiency (Sect. 25). Furthermore, the procedure of background subtraction ignores stray contributions such as side reception by the feed horn of a paraboloid; hence the integral measured over the main beam is less than the integral over all directions by a factor $\mathfrak{B}$, the beam ratio (Sect. 38). Thus, in terms of the T_a actually observed, we have for the absolute flux density of an extended source

$$S = \frac{2k}{\beta \lambda^2} \iint_{\text{beam}} T_a\, dx\, dy$$

where the beam efficiency β is given by

$$\beta = \mathfrak{B}\, \eta\,.$$

For the flux density of a point source it is sufficient to measure T_m the maximum value of T_a actually observed. Then from the relation $k\, T_m\, \Delta f = \frac{1}{2}\, S A\, \Delta f$ it follows that

$$S = \frac{2k\, T_m}{A}$$

$$= \frac{2k\, T_m}{\alpha\, \mathfrak{A}}$$

where the aperture efficiency α was shown in Sect. 38 to be given by

$$\alpha = \beta\, \mathfrak{D}\,.$$

b) Extended sources.

70. Pencil beam surveys. Consider first a moderately extended field, not too near the poles, and not too extensive in declination. Then the lines of constant right ascension and declination form a rectangular network.

Drift curves will be taken at declinations spaced one peculiar interval apart until the whole field is covered. Other ways of covering the field are possible but an essential character of the data is always preserved, viz. that one independent variable varies continuously and the other assumes discrete values.

It is necessary to reduce the data to sets of isophotes without introducing spurious detail which occasionally afflicts published surveys and with due regard to aerial smoothing. Now a drift curve, which is a cross-section of a function $T_a(x, y)$ with a cut-off (two-dimensional) spectrum, has itself a cut-off (one-dimensional) spectrum, and hence should be free from high-frequency components. However such components may be present because of errors or noise, and should be removed since they are unwarranted. Noise is best smoothed out by the eye but accurate filtering can also be carried out simply by convolution with $\dfrac{\sin \pi X}{\pi X}$, if X is the coordinate along the drift curve measured in units of one peculiar interval. Since X varies continuously the convolution integral does not reduce to a summation; but to compute the integral one would evaluate a summation,

$$T_\varepsilon = T_a * III\left(\frac{X}{\varepsilon}\right)\frac{\sin \pi X}{\pi X}$$

which would approach the desired integral

$$T_F = T_a * \frac{\sin \pi X}{\pi X}$$

as ε approached zero. Just how small ε need be we now enquire. Beginning with the coarsest interval, $\varepsilon = 1$, we find $T_1 = T_a$, that is, no filtering has been achieved; but with $\varepsilon = \tfrac{1}{2}$ we find

$$T_{\frac{1}{2}} = T_a * III(2X)\frac{\sin \pi X}{\pi X}$$

whence

$$\overline{T}_{\frac{1}{2}} = \overline{T}\left\{\frac{1}{4 s_c}III\left(\frac{s}{4 s_c}\right) * II\left(\frac{s}{2 s_c}\right)\right\}.$$

Hence $\overline{T}_{\frac{1}{2}}$ consists of a central part $II\left(\dfrac{s}{s_c}\right)\overline{T} = \overline{T}_F$ plus remoter parts. For many purposes this simple operation would be sufficient filtering, since the components to be rejected would often be chiefly just beyond the central region.

To recapitulate, T_a is filtered by reading off values at intervals of one half the peculiar interval and taking the convolution (more aptly serial product) with the series $\dfrac{\sin \pi X}{\pi X}$, where X runs through all half integers. This series is tabulated here, omitting the zeros and the value for $X = 0$.

X	$\dfrac{\sin \pi X}{\pi X}$	X	$\dfrac{\sin \pi X}{\pi X}$	X	$\dfrac{\sin \pi X}{\pi X}$	X	$\dfrac{\sin \pi X}{\pi X}$	X	$\dfrac{\sin \pi X}{\pi X}$
$\frac{1}{2}$	0.6366	$7\frac{1}{2}$	-0.0424	$14\frac{1}{2}$	0.0220	$21\frac{1}{2}$	-0.0148	$28\frac{1}{2}$	0.0112
$1\frac{1}{2}$	-0.2122	$8\frac{1}{2}$	0.0374	$15\frac{1}{2}$	-0.0205	$22\frac{1}{2}$	0.0141	$29\frac{1}{2}$	-0.0108
$2\frac{1}{2}$	0.1273	$9\frac{1}{2}$	-0.0335	$16\frac{1}{2}$	0.0193	$23\frac{1}{2}$	-0.0135	$30\frac{1}{2}$	0.0104
$3\frac{1}{2}$	-0.0909	$10\frac{1}{2}$	0.0303	$17\frac{1}{2}$	-0.0182	$24\frac{1}{2}$	0.0130	$31\frac{1}{2}$	-0.0101
$4\frac{1}{2}$	0.0707	$11\frac{1}{2}$	-0.0277	$18\frac{1}{2}$	0.0172	$25\frac{1}{2}$	-0.0125	$32\frac{1}{2}$	0.0098
$5\frac{1}{2}$	-0.0579	$12\frac{1}{2}$	0.0255	$19\frac{1}{2}$	-0.0163	$26\frac{1}{2}$	0.0120	$33\frac{1}{2}$	-0.0095
$6\frac{1}{2}$	0.0490	$13\frac{1}{2}$	-0.0236	$20\frac{1}{2}$	0.0155	$27\frac{1}{2}$	-0.0116	$34\frac{1}{2}$	0.0092

The test for adequacy of filtering is to compare T_a in the most unfavorable areas with the interpolated values discussed below. Where further filtering is required one repeats the first process with the same tabulated series but with $\varepsilon = \frac{1}{4}$.

There is an even simpler procedure which appears at first sight to be equivalent to filtering. If T_a is read off at discrete intervals corresponding to the desired cut-off, the set of values so obtained defines a function T' with cut-off spectrum. But this function is not the same as T_F since it is affected by high-frequency components of T_a. However, T' may often represent T_a adequately within the limits of accuracy of the observations.

The filtered data may be used to commence plotting points on isophotes. In areas where the isophotes are closely packed or highly curved it may be desirable to interpolate between the filtered drift curves. This is done by reading off a set of values along a transversal which intersects the drift lines. Convolution with the series tabulated above then gives an interpolated value.

When the field being surveyed is extended in declination the simplest procedure is to split it into smaller zones, and when the area is near the poles the drift curves may still be filtered and interpolation along hour circles is still possible. The apparent complication is considerably relieved by the use of a projection suited to the area being worked on.

Absolute values of sky brightness temperatures can be measured by taking account of losses as in Sect. 69. Suppose that a paraboloidal reflector points at an extended area of sky of brightness temperature T which fills the main beam (Sect. 38), thus eliminating considerations of aerial smoothing which will be returned to in Sect. 73. What will be the effective aerial temperature T_a actually observed? Power entering the aerial terminals distributes itself so that a fraction β is launched in the main beam (Sect. 38) and a fraction η all told is launched skywards (Sect. 25). Hence $1 - \eta$ is absorbed by the ground at temperature T_0. The remaining stray fraction $\eta - \beta$ proceeds skywards, but not in the main beam. By the principle of detailed balancing (Sect. 50) it follows that

$$T_a = \beta T + (1 - \eta) T_0 + (\eta - \beta) T_{\mathrm{av}},$$

where T_{av} is a weighted average of sky temperature over the directions taken by the stray radiation. Hence the temperature T, which it is desired to measure, is given in terms of the observed T_a by

$$T = \frac{T_a}{\beta} + \frac{1 - \eta}{\beta} T_0 + \frac{\eta - \beta}{\beta} T_{\mathrm{av}}.$$

The measurement of β and η has been discussed earlier. They are difficult to determine and may vary with aerial pointing. All the terms must, however, be estimated, and reference to the rich literature of sky surveys will reveal the wide variety of approaches that have been adopted.

71. Interferometric studies. From Sects. 55 and 56 it is clear that measurements of fringe visibility with an elementary interferometer at all spacings and in all azimuths give the Fourier transform of a source distribution and hence the distribution itself. The one dimensional form of this statement was given by McCready, Pawsey and Payne-Scott in their original paper on radio interferometric observations and the procedure has since been put into effect many times e.g. by Stanier[1] and Scheuer and Ryle[2] in one-dimension and by O'Brien[3] and Firor[4] in

[1] H.M. Stanier: Nature, Lond. **165**, 354 (1950).
[2] P.A.G. Scheuer and M. Ryle: Monthly Notices Roy. Astronom. Soc. London **113**, 3 (1953).
[3] P.A. O'Brien: Monthly Notices Roy. Astronom. Soc. London **113**, 597 (1953).
[4] J. Firor: Astrophys. Journ. **123**, 320 (1956).

two dimensions. The measurement of pha Γ is difficult and is not important in the case of symmetrical objects such as the quiet Sun except insofar as complete phase reversals occur. This has been investigated by allowing a small signal from a central aerial to leak into the transmission line in a way analogous to the determination of the phase in optical interference patterns by the admission of a little direct light.

In practice one can explore only that finite region of the transform plane corresponding to accessible interferometer spacings. Two-dimensional Fourier synthesis then yields a celestial distribution, which is not in general the true distribution, but the "principal solution" described below in Sect. 73. This procedure has been called aperture synthesis by RYLE[1]. When the elements of the interferometer are large the Fourier transform relation is modified (Sect. 55), but not greatly if, as is usually the case, the sources under study are compact relative to the beamwidth of a single element. When an interferometer has elements that are segmented or in any way complicated, it may be conveniently studied with the help of the two-dimensional spectral visibility island diagram (Sect. 65).

An important practical matter is to decide how many azimuths should be chosen and how many spacings. In the case of a bounded object such as the Sun, the two-dimensional sampling theorem shows that it suffices to sample the coherence at intervals of about 50 wavelengths as determined by the diameter of the radio Sun. It is not clear that radial symmetry in the observations is desirable, in fact it leads to redundancy at small spacings; it would be better to move the aerials along two perpendicular lines and arrange for the coherence to be sampled at points of a rectangular grid.

72. Fan beams and strip integration. It has not always been possible, during the rapidly expanding phase of observational radio astronomy, to bring to bear fine pencil beams or two-dimensional interferometric techniques, but important advances have been made with instruments having resolution in only one dimension. This was so with the early interferometric width measurements of solar noise storms and discrete sources. A long but narrow aperture is equivalent to an interferometer used at all spacings up to the maximum aperture but is speedier since it obtains the equivalent information in the one scan over the Sun. Nevertheless it is a highly deficient instrument since it receives on a knife-edge or fan beam and therefore confuses separate sources lying along the strip of sky lying in the beam. The purpose of this section is to show the limitations and possibilities of fan beams. They have been of undoubted importance and are likely to remain so, for at any stage of progress it is more feasible to improve resolution in one dimension than two, and experience has shown that new discoveries regarding detailed structure of emitting sources result from such partial improvement, though the effort of observation and reduction may then be too great to be suitable for the following phases of investigation.

Let $T(x, y)$ be a true distribution of brightness temperature and let a long and very narrow aerial receive radiation from a strip lying along the line

$$x \cos \vartheta + y \sin \vartheta - R = 0.$$

Then we distinguish between line integrating and strip integrating according as the strip is of infinitesimal width or is widened out with a profile A corresponding to the long though finite dimension of the aerial. The line integral of $T(x, y)$ is defined by

$$T_L(R, \vartheta) = \iint T(x, y)\, \delta(x \cos \vartheta + y \sin \vartheta - R)\, dx\, dy$$

[1] M. RYLE: Nature, Lond. **180**, 110 (1957).

and the strip integral by

$$T_S(R, \vartheta) = \iint T(x, y) A(x \cos \vartheta + y \sin \vartheta - R) \, dx \, dy.$$

If we have $T_L(R, \vartheta)$ for $\vartheta = \vartheta_1$ only, an important case in practice, the integral equation is not soluble; for if $T_1(x, y)$ satisfies the line integral equation when $\vartheta = \vartheta_1$, so also does $T_1(x, y) + T_2(x, y)$, where $T_2(x, y)$ is an invisible distribution, that is one such that

$$\iint T_2(x, y) \, \delta(x \cos \vartheta + y \sin \vartheta_1 - R) \, dx \, dy = 0.$$

If, however, in addition to $T_L(R, \vartheta_1)$ we have other information about the source distribution, then further progress may be made. For example if it is known that the source has circular symmetry then the problem may be fully solved, as explained below.

Now the strip integral equation will suffer from invisible distributions of the kind discussed in earlier sections, for $T_S(R, \vartheta_1)$ will differ from $T_L(R, \vartheta_1)$ by the absence of Fourier components beyond a cut-off set by the profile A. Hence $T_S(R, \vartheta)$, regarded as a two-dimensional distribution in the xy-plane, will have a Fourier transform in the uv-plane which does not extend beyond a central circular island, points on whose perimeter represent spatial frequencies equal to the cut-off value. It is thus sufficient to study the line integration problem here, i.e. the confusion of information, and to study the question of the loss of information and accompanying effects due to the cut-off in later sections on restoration.

In the case of a source with circular symmetry which has been scanned in one direction $\vartheta_1 = 0$ with a line beam, we have, writing $T_L(x)$ for $T_L(R, 0)$,

$$T_L(x) = \iint T(\sqrt{x^2 + y^2}) \, \delta(x - R) \, dx \, dy$$

$$= 2 \int\limits_{r=x}^{\infty} T(r) \, dy,$$

where

$$r^2 = x^2 + y^2,$$

and so

$$T_L(x) = 2 \int\limits_{x}^{\infty} \frac{T(r) \, r \, dr}{\sqrt{r^2 - x^2}}.$$

Substituting $\xi = x^2$ and $\varrho = r^2$ and writing $T_L(x) = \hat{T}_L(x^2)$, $T(r) = \hat{T}(r^2)$,

$$\hat{T}_L(\xi) = \int\limits_{-\infty}^{\infty} K(\xi - \varrho) \, \hat{T}(\varrho) \, d\varrho,$$

or

$$\hat{T}_L = K * \hat{T}$$

where

$$K(\xi) = \begin{cases} (-\xi)^{-\frac{1}{2}} & (\xi < 0), \\ 0 & (\xi \geqq 0). \end{cases}$$

In this form as a convolution integral the equation is soluble, but the solution, $\pi \hat{T} = -K * \hat{T}_L'$, is not as convenient numerically as the direct inversion of the process $K * \hat{T}$, starting from the outlying parts where $\hat{T}$ is zero and working inwards.

In graphical terms (Fig. 54) one evaluates the line integral along AB over the shaded outer area where T is known, and deduces the next inner value from the correction necessary to give the observed full integral. A graphical generalization to the case of no symmetry is clearly possible, and has been discussed by BRACEWELL[1] in a paper which also presents a universal table of coefficients for the rapid inversion from $\widehat{T_L}$ to $\widehat{T}$.

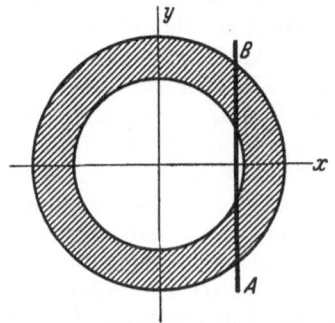

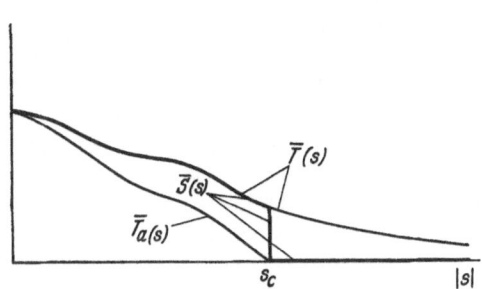

Fig. 54. Illustrating the inversion of line integration.   Fig. 55. The relation between $\overline{T}_a(s)$ and $\overline{T}(s)$.

c) Restoration.

73. The principal solution. It was shown in Sect. 53 that the normalized response of an aerial to a sinusoidal component of brightness falls with increasing spatial frequency s from unity, for a uniform distribution, to zero for a spatial frequency s_c cycles per radian, where the extent of the aerial is s_c wavelengths. Fig. 55 shows the relation between the transform of the observed distribution $\overline{T}_a(s)$ and that of the true distribution $\overline{T}(s)$. For the purposes of the figure, a one dimensional case has been illustrated in which $\overline{T}_a$ and $\overline{T}$ are real and even functions of s.

In Sect. 51 it was shown that there are an infinite number of solutions of the equation

$$T_a = A * T;$$

however, T_a itself is not one of them, and the question arises what to do to find a distribution which could have given rise to the observed T_a.

If there is no knowledge of $T(x)$ other than that contained in $T_a(x)$, the most that can be done is to restore to their full value those (low-frequency) components of T which, while present in T_a, have been reduced in amplitude. This gives a unique result, which is a solution of the equation and which is called the principal solution, $S(x)$. It may be defined as that solution whose transform is the same as the transform of the true distribution at all values of s for which $\overline{A}(s) \neq 0$, and zero elsewhere, that is,

$$\overline{S}(s) = \begin{cases} \dfrac{\overline{T}_a(s)}{\overline{A}(s)}, & (\overline{A}(s) \neq 0) \\ 0, & (\overline{A}(s) = 0). \end{cases}$$

The principal solution has the distinction that, among all the approximate solutions not containing components beyond s_c, it is the best least-mean-squares fit to the true distribution. For if $S(x) + H(x)$ is one of the said approximations

[1] R.N. BRACEWELL: Austral. J. Phys. **9**, 198 (1956).

then

$$\int_{-\infty}^{\infty} |T(x) - \{S(x) + H(x)\}|^2\, dx = \int_{-\infty}^{\infty} |\overline{T} - (\overline{S} + \overline{H})|^2\, ds = \int_{-s_c}^{s_c} |\overline{H}|^2\, ds + \text{const},$$

which is a minimum when $\overline{H} = H = 0$.

When there are no errors, the information contained in S is exactly the same as that given by an interferometer consisting of two isotropic aerials used at all spacings from zero up to the full aerial width. However the interferometer gives the Fourier components of T directly at their full value, whereas an aerial consisting of a single aperture weights the components so that they must subsequently be restored. As components near the cut-off frequency receive little weight the error spectrum near the cut off is important and it has been felt that the application of a large compensating factor would be deleterious. Whilst this is true it does not follow that the variable-spacing interferometer gives a superior result, for the low weight assigned to the extreme components as received by the single aperture is that weight which is assigned evenly to the components observed interferometrically. The single aperture emphasizes the components of low spatial frequency, and restoration may be regarded as de-emphasis of the superabundant. This will be clear if a comparison is made between a Christiansen array of large paraboloids and a two element interferometer using a pair of the same paraboloids. But in this case the spectral sensitivity function does not fall uniformly away to zero, and the compensating factor does not become infinite. The case of a uniform aperture may appear to be different but here also a comparison may be made with the aerials proposed for the interferometer. Since the effective area of an aerial cannot fall below $\frac{1}{8}\lambda^2$, an apparently uniform aperture is equivalent to an array of elementary dipoles spaced 8 per square wavelength. Its spectral sensitivity function does not, strictly speaking, descend regularly to zero, and the compensating factor should never exceed the limit set by this consideration. Even so, full restoration may be deleterious, if a single dipole is insufficient on its own. It then leads to spurious detail to de-emphasize the components of low spatial frequency to the same state. The problem of optimum restoration in the presence of errors has been discussed[1]; the best spectral restoring factor, when there is no correlation between the error distribution and T_a, is

$$\frac{1}{\overline{A}\left[1 + \dfrac{\langle \overline{E}\,\overline{E}^* \rangle}{\overline{T}_a\,\overline{T}_a^*}\right]}$$

where $\langle \overline{E}\,\overline{E}^* \rangle$ is the ensemble average squared modulus of the error distribution.

The degree of approximation of the principal solution to the true distribution depends markedly on the form of T. Thus when T contains no spectral components at those frequencies where $\overline{A}(s) = 0$, S is identical with T, and when T has components only where $\overline{A}(s)$ is not very different from unity, restoration may be carried out with confidence. However, when the spectrum of T is still appreciable at the cut-off frequency s_c, the resulting discontinuity in the spectrum of the principal solution can cause spurious oscillations in S. This may result in S assuming negative values which in radio astronomy is impossible.

There are various ways of dealing with the discontinuity. In an analogous problem in X-ray crystallography[2] it is smoothed out arbitrarily. A little cautious extrapolation may be resorted to. A better approach seems to be to seek

[1] R. N. Bracewell: Proc. Inst. Radio Engrs. **46**, 106 (1958).

[2] J. Waser and V. Schomaker: Rev. Mod. Phys. **25**, 671 (1953).

some physical theory which predicts a form for T, or to adopt some multi-parameter representation which is physically acceptable, and then evaluate the parameters by use of the finite number of independent values of $\overline{T}$. This is an extension of the procedures for deducing the extent of a small source from a measurement of one value of $\overline{T}$, where a one-parameter family of shapes was fitted to the observation.

74. The method of successive substitutions. We may now consider actual methods of restoration in the light of the above. Suppose that T_{app} is some approximation to T. Then it can be tested by scanning with the aerial pattern and comparing with T_a. The discrepancy $T_a - A * T_{\mathrm{app}}$ is taken as a first estimate of the difference between T_{app} and T. This leads to a further approximation

$$T_{\mathrm{app}} + (T_a - A * T_{\mathrm{app}}).$$

If T_a itself is taken as the initial approximation to T, one obtains as the first approximate restoration

$$T_1 = T_a + (T_a - A * T_a).$$

By applying the same procedure to T_1 we have a second restoration

$$T_2 = T_1 + (T_a - A * T_1)$$

and so on. In practice the iteration is halted when smoothing the trial distribution with A gives a result agreeing with T_a within the experimental error. The method has been studied in detail by BRACEWELL and ROBERTS[1] who found the condition for convergence of the sequence to be that $|1 - \overline{A}(s)| < 1$ for all s such that $\overline{T}_a(s) \neq 0$, and showed that the limit of the sequence is the principal solution $S(x)$.

This method is feasible in practice and has been often used. However, it is laborious in two dimensions and probably will not be used in current high resolution studies except for special cases. A new approach to cases where there is much intricate detail is to admit the tentative character even of the principal solution and to seek methods of modest accuracy but which are simple to apply.

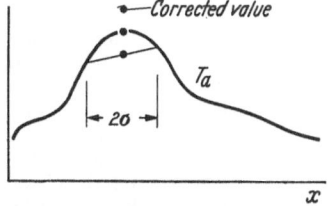

Fig. 56. The chord construction for restoration.

75. The chord contruction. An example of a simple method is afforded by the chord construction and its generalization to two dimensions[2]. In Fig. 56 the corrected value lies above the observed curve $T_a(x)$ by as much as the curve lies above the midpoint of a chord of constants pan 2σ, centered at the point in question. The quantity σ is the standard deviation of the instrumental profile in general, but in radio astronomy where σ is never finite, the span is determined by a matching technique. In two dimensions it is useful to imagine a plane corresponding to the chord, and the correction at any point, determined from the four neighbouring values, is

$$-\tfrac{1}{4}(\Delta_{xx} + \Delta_{yy})\, T_a,$$

where $\Delta_{xx} T_a$ is the second difference of T_a when y is kept constant, and the interval over which the differencing is done is equal to $\sqrt{2}$ standard deviations for

[1] R.N. BRACEWELL and J.A. ROBERTS: Austral. J. Phys. **7**, 615 (1954).

[2] R.N. BRACEWELL: J. Opt. Soc. Amer. **45**, 873 (1955). — Austral. J. Phys. **8**, 54, 200 (1955).

a Gaussian beam. For beams other than Gaussian the correction becomes

$$- \left(\chi \, \Delta_{xx} + \psi \, \Delta_{yy} \right) T_a,$$

where χ, ψ and the differencing intervals α and β are fixed by matching $\chi \sin^2 \pi \alpha u + \psi \sin^2 \pi \beta v$ to $(\overline{A})^{-1} - 1$. The chord construction is equivalent to three stages of successive substitution.

Acknowledgment. This contribution contains material which was developed in connection with research in radio astronomy supported at Stanford University by the Office of Scientific Research of the United States Air Force.

Bibliography.

General.

Two books devoted entirely to radio astronomy, including techniques, are

Brown, R. H. and A. C. B. Lovell: The Exploration of Space by Radio. London 1951.
Pawsey, J.L., and R.N. Bracewell: Radio Astronomy, Oxford 1954.

The Proceedings of the Institute of Radio Engineers, Vol. 46, No. 1, 1958, ed. by F. T. Haddock is devoted to papers on radio astronomy which are very largely concerned with techniques.

Receivers.

Texts concerned with radar receivers are a source of information relative to receivers for radio astronomy. Two very detailed books are

Valley, G.E., and H. Wallman: Vacuum Tube Amplifiers, New York 1948,
Voorhis, S.N. van: Microwave Receivers, New York 1948,

and a more general book with valuable chapters on receiver practice is

Bowen, E.G.: A Textbook of Radar, 2nd ed., Cambridge 1954,

which also contains an excellent chapter on aerials. A rich variety of microwave techniques are presented by E. L. Ginzton, Microwave Measurements. New York 1958.

Aerials.

There is a wide range of literature and it is still increasing. There are no books on aerials for radio astronomy, but among books with chapters relevant to radio astronomy are the following:

Silver, S.: Microwave Antenna Theory and Design. New York 1949. (This book is useful for diffraction theory of apertures and detailed discussion of parabolic reflectors.)
Smith, R.A.: Aerials for Metre and Decimetre Wavelengths. Cambridge 1949.
Fry, D.W., and F.K. Goward: Aerials for Centimetre Wavelengths. Cambridge 1950.
Schelkunoff, S.A., and H.T. Friis: Antennas Theory and Practice. New York 1952. (An undergraduate text concentrating on a logical development of relevant electromagnetic theory.)
Kraus, J.D.: Antennas. New York 1950. (An undergraduate text with much information on helical aerials.)

Aerial Smoothing.

Much of this subject will be found in volumes 7 to 9 of the Australian Journal of Physics. See **7**, 615 (1954); **8**, 54, 200 (1955); **9**, 198, 297 (1956) and Proc. Inst. Radio Engrs. **46**, 106 (1958).

Interferometers.

There are no books on radio interferometers, which so far have been discussed only by writers on radio astronomy. See Brown and Lovell, Pawsey and Bracewell, M. Ryle: Proc. Roy. Soc. Lond., Ser. A **211**, 351 (1952) and R. N. Bracewell: Proc. Inst. Radio Engrs. **46**, 97 (1958).

For early instrumental papers in the Russian language see V.V. Vitkevich: Dokl. Akad. Nauk. SSSR. **86**, 39 (1952); **91**, 1301 (1953); **102**, 469 (1955); Astronom. Zhurn. **29**, 450 (1952); (with R. L. Sorochenko) **30**, 631 (1953); **34**, 217 (1957), and Transactions of the Fifth Conference on Questions of Cosmogony, Moscow 1956.

Diffraction theory.

A good deal of the theory underlying the observational techniques of radio astronomy applies also in other fields. For related reading in optics see

MARÉCHAL, A.: Handbuch der Physik, Bd. XXIV, S. 44, 1956.
FRANÇON, M.: Handbuch der Physik, Bd. XXIV, S. 171, 1956.
BORN, M., and E. WOLF: Principles of Optics, London 1958,
WOLF, E.: Proc. Roy. Soc. Lond., Ser. A **225**, 96 (1954); **230**, 246 (1955),
and in ionospheric physics,
RATCLIFFE, J.A.: Rep. Progr. Phys. **19**, 188 (1956).

Information theory, noise, and Fourier theory.

An extensive body of theory exists which continually arises in connection with all the subjects of this chapter. For an interesting combination of topics of this general kind see

WOODWARD, P.M.: Probability and Information Theory, with Applications to Radar. London 1953.
BRILLOUIN, L.: Science and Information Theory. New York 1956.

The classical papers on fluctuation theory from the standpoint of electric circuits are those of RICE, especially those from volumes 23 and 24 of the Bell System Technical Journal which may be found reprinted in

WAX, N.: Noise and Stochastic Processes. New York 1954.

Bibliography.

The literature of radio astronomy is widely scattered in journals devoted to physics, astronomy, and electrical engineering. Adequate abstracting is provided by Electronic and Radio Engineer Abstracts, which are reprinted monthly in the Proceedings of the Institute of Radio Engineers, especially in the section on Geophysics and Extraterrestrial Phenomena, but the sections devoted to aerials, wave propagation, reception, and others, are also highly relevant. The astronomy section of Science Abstracts is another source. A very competent bibliography prepared by MARTHA STAHR CARPENTER has also been issued at intervals by Cornell University.